Cyber Investigations

Cyber Investigations

A Research Based Introduction for Advanced Studies

Edited by

André Årnes
Norwegian University of Science and Technology (NTNU)

Registered Offices
John Wiley & Sons, Inc., 111 River Street, Hoboken, NJ 07030, USA
John Wiley & Sons Ltd, The Atrium, Southern Gate, Chichester, West Sussex, PO19 8SQ, UK

Editorial Office
The Atrium, Southern Gate, Chichester, West Sussex, PO19 8SQ, UK

For details of our global editorial offices, customer services, and more information about Wiley products visit us at www.wiley.com.

Wiley also publishes its books in a variety of electronic formats and by print-on-demand. Some content that appears in standard print versions of this book may not be available in other formats.

Library of Congress Cataloging-in-Publication Data applied for:
Paperback ISBN: 9781119582311

Cover Design: Wiley
Cover Images: Background - © Mike Pellinni/Shutterstock; Left-hand image - Courtesy of Lasse Øverlier; Right-hand image - Courtesy of Jens-Petter Sandvik

Set in 10/12pt Warnock by Straive, Pondicherry, India
Printed and bound by CPI Group (UK) Ltd, Croydon, CR0 4YY

C9781119582311_031022

Contents

Preface

Dear cyber investigation student. You are holding in your hand the "Cyber Investigation" textbook written by faculty, associates, and former students of cyber security and digital forensics at the Norwegian University of Science and Technology (NTNU). The book is a sequel to our previous textbook on "Digital Forensics," and it represents our shared philosophy of learning cyber investigations. The book covers both technical, legal, and process aspects of cyber investigations, and it is intended for advanced and graduate students of cyber investigations, digital forensics, and cyber security. It is based on research, teaching material, experiences, and student feedback over several years.

The reason for embarking on this project is that there was no literature currently available within the area suitable as a stand-alone curriculum at an academic level, as most of the available literature is primarily intended for practitioners and technical readers. Consequently, literature tailored to academic education in cyber investigations is needed. As you can probably imagine, writing a textbook is a daunting task. While the authors have put much effort into making this version readable and easily available, we are keen to hear your feedback so that we can improve our teaching material over time.

I would like to thank the chapter authors for their dedicated and collaborative efforts to this project, as well as Professor Katrin Franke of the NTNU Digital Forensics Research Group. We are grateful for the support provided by the Norwegian Research Council through the ArsForensica project (project number 248094), the NTNU Center for Cyber and Information Security (CCIS), the Norwegian Police Directorate, Telenor Group, and the U.S. Embassy in Norway grant (grant number SNO60017IN0047) awarded by the U.S. State Department toward this work.

Good luck with learning Cyber Investigations!

André Årnes
Norway, May 2022

Companion Website

The figures and tables from this book are available for Instructors at:

http://www.wiley.com/go/cyber

List of Contributors

André Årnes, PhD, Siv.ing. (MSc), BA – Oslo, Norway
Professor, Norwegian University of Science and Technology (NTNU) and Partner & Co-owner White Label Consultancy, Oslo, Norway

- PhD and MSc in information security from NTNU, visiting researcher at UCSB, USA and Queens's University, Canada
- White Label Consultancy 2022–: Partner and Co-owner, with responsibility for cyber security
- Telenor 2010–2022: SVP and Chief Security Officer (from 2015 to 2022), CIO Global Shared Services (from 2013 to 2015)
- National Criminal Investigation Service (Kripos) 2003–2007: Special Investigator within computer crime and digital forensics
- GIAC Certified Forensic Analyst (GCFA), IEEE Senior Member, and member of the Europol Cyber Crime Centre (EC3) Advisory Group for communications providers.

Nina Sunde, MSc – Oslo, Norway
Police Superintendent, Lecturer at Department for Post Graduate Education and Training, The Norwegian Police University College, Oslo, Norway

- PhD student in Cybercrime Investigations, Faculty of Law, University of Oslo (2018–)
- MSc in Information Security and Cybercrime Investigation, NTNU (2017)
- Lecturer at The Norwegian Police University College (2012–current)
- Police Superintendent at National Criminal Investigation Service (Kripos), investigation of cybercrime (2003–2009)
- Permanent member at European Working Party on Information Technology Crime (EWPITC), Interpol (2007–2010)
- Police Detective, Department for Investigation of Homicide, The Oslo Police District (2001–2003).

Inger Marie Sunde, PhD, LLM, BA – Oslo, Norway
Professor, The Norwegian Police University College, Oslo, Norway

- The Norwegian Police University College *2010–current*: Professor (from 2014)
- The Strategic Group on Ethics, reporting to the European Clearing Board of Innovation in the Europol Innovation Lab. Co-leader. 2022–
- Bergen University, The Police and Prosecution Project *2019–2022:* Professor II
- The Norwegian Parliamentary Oversight Committee on Intelligence and Security Services – 2014–2019
- The Norwegian Defense University College *2004:* Chief's Main Study
- Senior Public Prosecutor; Head of the Norwegian Cybercrime Center; Visiting researcher Max Planck Institute; PhD University in Oslo; Founder and leader of the Research Group "Police and Technology"
- Author and editor of publications on cybercrime law

Petter Christian Bjelland, MSc – Oslo, Norway
Digital Forensics and Cyber Investigations Expert, Oslo, Norway

- Manager Digital Forensics, EY Norway (2017–2018)
- Advisor Digital Investigations at the National Criminal Investigation Service Kripos (2015–2017)
- Senior Software Engineer in the Norwegian Defense (2011–2015)
- MSc in digital forensics from Gjøvik University College 2014
- Peer-reviewed paper at DFRWS Europa 2014 and in Elsevier Digital Investigation.

Jens-Petter Sandvik, Cand. Scient. – Oslo, Norway
Senior Engineer in Digital Forensics, National Cybercrime Center/NC3, National Criminal Investigation Service/ Kripos, Oslo, Norway and Department of Information Security and Communication Technology, Faculty of Information Technology and Electrical Engineering, Norwegian University of Science and Technology (NTNU), Trondheim, Norway

- PhD student at NTNU Digital Forensics Laboratory (2017–*current*)
- Senior Engineer in Digital Forensics, Kripos, the Norwegian Criminal Investigation Service (2006–*current*)
- Software Developer Malware Detection, Norman ASA (2001–2005)
- Cand. Scient., University of Oslo 2005.

Lasse Øverlier, PhD, Siv.ing. (MSc), MTM – Trondheim, Norway

Associate Professor, Department of Information Security and Communication Technology, Faculty of Information Technology and Electrical Engineering, Norwegian University of Science and Technology (NTNU), Trondheim, Norway and Principal Scientist, Norwegian Defence Research Establishment (FFI), Gjøvik, Norway

- PhD Information Security, University of Oslo 2007
- Associate Professor, NTNU (2002–*current*)
- Principal Scientist, Norwegian Defence Research Establishment (FFI) (2002–*current*)
- Research Scientist, Army Research Laboratory, California/Maryland (2015–2016)
- Research Scientist, Naval Research Laboratory, Washington, DC (2005–2006)
- Co-founder and technical manager EUnet Media AS (later KPNQwest).

Kyle Porter, PhD, MSc – Gjøvik, Norway

Researcher, Department of Information Security and Communication Technology, Norwegian University of Science and Technology (NTNU), Gjøvik, Norway

- PhD in Information Security with a focus in Digital Forensics, NTNU (2017–2022)
- MSc in Information Security, NTNU (2017)
- BA in Mathematics, University of Washington (2012)
- Author of several scientific papers.

List of Figures

List of Tables

List of Examples

List of Definitions

List of Legal Provisions

List of Equations

List of Abbreviations

5WH	Who, Where, What, When, Why, and How
6LoWPAN	IPv6 over Low-power Wireless Personal Area Network
ABC	Assume nothing, Believe nothing, Challenge everything
ACM	Association for Computing Machinery
ACPO	Association of Chief Police Officers
ADC	Analog-to-Digital Converter
AFF	Advanced Forensics File format
AI	Artificial Intelligence
ANB	Analyst's Notebook
API	Application Programming Interface
APT	Advanced Persistent Threat
ATT&CK	Adversarial Tactics, Techniques, and Common Knowledge
BGP	Border Gateway Protocol
BSI	German Federal Office for Information Security
C2	Command and Control
CCD	Charge-Coupled Device
CETS	Council of Europe Treaty Series
CFA	Color Filter Array
CFREU	Charter of Fundamental Rights of the European Union
CIA	Central Intelligence Agency
CIDR	Classless Inter-Domain Routing
CIS	Center for Internet Security
CMOS	Complementary Metal-Oxide Semiconductor
CNN	Convolutional Neural Network
CoAP	Constrained Application Protocol
CPS	Cyber-Physical System
DDoS	Distributed Denial-of-Service
DFRWS	Digital Forensics Research Workshop

DHCP	Dynamic Host Configuration Protocol
DNS	Domain Name System
ECHR	European Court of Human Rights
EEC	European Economic Community
EM	Expectation-Maximization
ENF	Electric Network Frequency
ENFSI	European Network of Forensic Science Institutes
EU	European Union
EUCFR	European Union Charter of Fundamental Rights
EWF	Expert Witness Format
EXIF	Exchangeable Image File Format
FAIoT	Forensic-Aware IoT
FBI	United States Federal Bureau of Investigation
FEMS	Forensic Edge Management System
FISA	Foreign Intelligence Surveillance Act
FKIE	Fraunhofer Institute for Communication, Information Processing, and Ergonomics
FTP	File Transfer Protocol
GAN	Generative Adversarial Network
GDPR	General Data Protection Regulation
GIAC	Global Information Assurance Certification
GNSS	Global Navigation Satellite System
GOP	Group of Pictures
GPS	Global Positioning System
HDFS	Hadoop Distributed File System
HEVC	High-Efficiency Video Coding
HTTP	HyperText Transfer Protocol
ICANN	Internet Corporation for Assigned Names and Numbers
ICCPR	International Covenant on Civil and Political Rights
ICIP	Integrated Cyber Investigation Process
ICJ	International Court of Justice
ICMP	Internet Control Message Protocol
IDS	Intrusion Detection System
IEC	International Electrotechnical Commission
IEEE	Institute of Electrical and Electronics Engineers
IETF	Internet Engineering Task Force
IFTTT	If This Then That
IIoT	Industrial Internet of Things

IMAP	Internet Message Access Protocol
IoT	Internet of Things
IP	Internet Protocol
IPTC	International Press Telecommunications Council
IPv6	Internet Protocol version 6
ISO	International Organization for Standardization
ISP	Internet Service Provider
ITS	Intelligent Transportation System
JAP	Java Anon Proxy
JIT	Joint Investigation Team
JPEG	Joint Photographic Experts Group
LAN	Local Area Network
LE	Law Enforcement
LEA	Law Enforcement Agency
LED	Law Enforcement Directive (European Union)
LoWPAN	Low-power Wireless Personal Area Network
LPWAN	Low-Power Wide Area Network
LTE	Long-Term Evolution
M2M	Machine-to-Machine
MAC	Media Access Protocol
MB	Megabyte
MCH	Multiple Competing Hypotheses
NB-IoT	Narrow-Band IoT
NER	Named Entity Recognition
NIS	National Intelligence Services
NIST	National Institute of Standards and Technology
NSA	United States National Security Agency
NTLK	Natural Language Toolkit
OCR	Object Character Recognition
OMP	Ortholinear Matching Pursuit
OSINT	Open-Source Intelligence
P^5	Peer-to-Peer Personal Privacy Protocol
PCI	Private sector Investigator
PDF	Portable Document Format
PET	Privacy Enhancing Technologies
PFS	Perfect Forward Secrecy
PGP	Pretty Good Privacy
PI	Private Investigator

PRNU	Photoresponse Non-uniformity
RFC	Request for Comments
RFID	Radio Frequency Identification
ROC	Receiver Operating Characteristic
RPL	IPv6 Routing Protocol for Low-power and lossy networks
SANS	SysAdmin, Audit, Network, and Security Institute
SDN	Software Defined Network
SIFT	Scale-Invariant Feature Transform
SMB	Server Message Block
SNA	Social Network Analysis
SNR	Signal to Noise Ratio
SOA	Service Oriented Architecture
SOCMINT	Social Media Intelligence
SVM	Support Vector Machine
TB	Terrabyte
TCP	Transmission Control Protocol
TFC	Traffic Flow Confidentiality
TLS	Transport Layer Security
TM	Tallinn Manual 2.0
UDP	User Datagram Protocol
VANET	Vehicular Ad Hoc Network
VPN	Virtual Private Network
WPAN	Wireless Personal Area Network
WSN	Wireless Sensor Network
XMP	Extensible Metadata Platform

1

Introduction

André Årnes[1,2,] *

[1] *White Label Consultancy, Oslo, Norway*
[2] *Norwegian University of Science and Technology (NTNU), Oslo, Norway*

As of 2021, there are more than 5 billion Internet users globally, representing close to 60% of the world's population (Internet World Stats, 2021). The growth of the Internet is coupled with an estimated 35 billion Internet of Things (IoT) connections as of 2020, expected to grow by an estimated 20 billion connected devices, expected to grow by 180% by 2024 (Juniper Research, 2020). As a result, digital services are becoming central to and often a necessity in criminal investigations, ranging from traditional evidence like telecommunication call data records and location data and financial transactions to the comprehensive personal tracking of Google timeline, personal health trackers like Fitbit, connected cars, bitcoin transactions. The criminal system is *drowning in digital evidence* (Burgess, 2018).

1.1   Introduction

One of the earliest public accounts of cybercrime and cyber investigations, "Crime by Computer," was given by Donn B. Parker as early as 1976 (Parker, 1976), documenting a wide range of *"startling new kinds of million-dollar fraud, theft, larceny & embezzlement."* The widely known firsthand accounts of cyber investigations told by Cliff Stoll in "The Cuckoo's Egg: Tracking a Spy Through the Maze of Computer Espionage" (Stoll, 1989) has later set the standard for hunting hackers, and the author remains a cybersecurity icon (Greenberg, 2019). Throughout this story, Stoll takes us through a detective story in its own right, where he detects, investigates, and documents evidence of cyber espionage. In another classic paper based on an actual event from 1991, "An Evening with Berferd – In Which a Cracker is Lured, Endured, and Studied," Cheswick

---

* Professor André Årnes is a Partner and Co-Owner of White Label Consultancy. He has served as the Global Chief Security Officer of Telenor Group from 2015 to 2022, and he was a Special Investigator with the Norwegian Criminal Investigation Service (Kripos) from 2003 to 2008. He is a part-time Professor at Norwegian University of Science and Technology within cyber security and digital forensics.

(1997) of the AT&T Bell laboratories retells the story of how they followed a hacker for months to trace his location and learn his techniques.

Even in these early stories, we learned that cyber investigations are complex processes that require dedication, persistence, and efforts, a strong sense of curiosity, as well as expert competencies. There is a call for a systematic approach, like the cyber investigation process in this book. A central topic in any cyber investigation will be digital forensics, which was the topic of our previous textbook (Årnes, 2018).

As described in Årnes (2018), an investigation is a systematic examination to identify or verify facts. A key objective during an investigation is to determine facts related to a crime or incident. A standard methodology is the 5WH model, which defines the goals of an investigation as Who, Where, What, When, Why, and How (Stelfox, 2013; Tilstone et al., 2013). In this book, we define an investigation as follows:

Definition 1.1: Investigation

A systematic collection, examination, and evaluation of all information relevant to establish the facts about an incident or an alleged crime and identify who may have committed or contributed to it.

1.2 Cybercrime and Cybersecurity

Cybercrime generates around 1.5 trillion USD per year, and global damages are estimated to be more than 6 trillion per year by 2021. Almost 700 million people are victims of cybercrime, and businesses take an average of 196 days to detect an attack. The rate and broad impact of successful attacks lead to claims that the system is broken and calls for new and more intelligent measures (Gault, 2015). In Europol's annual cybercrime report (Europol, 2019), Europol asserts that *"Cybercrime is continuing to mature and becoming more and more bold, shifting its focus to larger and more profitable targets as well as new technologies. Data is the key element in cybercrime, both from a crime and an investigate perspective."*

1.2.1 Cybercrime

As society increasingly depends on our digital infrastructure, the potential for digital abuse, criminal activity, and even warfare on the Internet increases. *Cybercrime* is a frequently used term that refers to both crimes targeting the Internet itself and activities that utilize the Internet to perform a crime. In this book, we use the following simple definition:

Definition 1.2: Cybercrime

"Crime or illegal activity that is done using the Internet." (Cambridge Dictionary, 2022)

For an additional perspective, cybercrime was defined as either advanced cybercrime or cyber-enabled crime in a publication by Interpol (2018):

- *Advanced cybercrime* (*or high-tech crime*): "sophisticated attacks against computer hardware and software."
- *Cyber-enabled crime*: "traditional crimes that have taken a new turn with the advent of the Internet, such as crimes against children, financial crimes, and terrorism."

1.2.2   Cybercriminals and Threat Actors

To understand cybercrime, one needs to understand the cybercriminals, or threat actors, which is the common reference in cybersecurity. A threat actor is an actor in cyberspace that performs malicious or hostile activities. There are many categorizations, and for this book, we will depend on the definitions by RAND Corporation and the Canadian Centre for Cybersecurity as discussed below.

In a testimony presented before the House Financial Services Committee, Subcommittee on Terrorism and Illicit Finance, on March 15, 2018, the RAND Corporation classified the threat actors as follows (Ablon, 2018):

— Cyberterrorists
— Hacktivists
— State-sponsored Actors
— Cybercriminals

In a more extensive list intended for public use, the Canadian Centre for Cyber Security (2018) classifies the threat actors as follows:

— Nation States
— Cybercriminals
— Hacktivists
— Terrorist Groups
— Thrill Seekers
— Insider Threats

Based on the models outlined above, we will adopt a three-tier cyber threat actor (CTA) model in this book, distinguished through the resources available and level of organization, often referred to as a *threat actor pyramid*:

- *National*: Nation-states
- *Organized*: Cybercriminals, hacktivists, and terrorist groups
- *Individual*: Thrill seekers and insider threats

Definition 1.3: Cyber threat actor

"A CTA is a participant (person or group) in an action or process that is characterized by malice or hostile action (intending harm) using computers, devices, systems, or networks." (Center for Cyber Security (CIS), 2019).

1.2.3 Cybersecurity

To protect our digital infrastructure against cybercrime, we depend on cybersecurity or measures to protect a computer or computer system against unauthorized access or attack. As a result, cybersecurity has become a rapidly growing industry, as society is scrambling to protect our rapidly developing technology against increasingly advanced cybercriminals.

Definition 1.4: Cybersecurity
"Measures taken to protect a computer or computer system (as on the Internet) against unauthorized access or attack." (Marriot Webster, 2022)

1.2.4 Threat Modeling – Cyber Kill Chain and MITRE ATT&CK

To support cybersecurity efforts against nation-state threat actors (so-called APTs – advanced persistent threats), Lockheed Martin developed the Cyber Kill Chain (Hutchins, 2011), which defined the stages of an attack starting with reconnaissance and resulting in actions on objectives. The term "kill chain" refers to a military term describing the structure of an attack. The Cyber Kill Chain defines the following phases of a successful attack:

1) Reconnaissance
2) Weaponization
3) Delivery
4) Exploitation
5) Installation
6) Command and Control (C2)
7) Actions on Objectives

While the Cyber Kill Chain provided an increased understanding of the anatomy of cyberattacks, the concept needed to be developed further to perform threat modeling and threat assessments effectively. *MITRE ATT&CK™* (MITRE, 2022) is a comprehensive description of cyber attackers' behavior once inside and embedded in a computer network. The model is based on publicly known adversarial behavior and is continuously updated. The highest level of abstraction, referred to as *tactics* represent the adversary's tactical goals, are:

- *Reconnaissance*
- *Resource development*
- *Initial access*
- *Execution*
- *Persistence*
- *Privilege escalation*
- *Defense evasion*
- *Credential access*
- *Discovery*
- *Lateral movement*

- *Collection*
- *Command and control*
- *Exfiltration*
- *Impact*

1.3   Cyber Investigations

The investigation of cybercrime (a criminal event) or a cybersecurity incident (a breach of security) can be referred to as a cyber investigation or a cybercrime investigation. The purpose of a cyber investigation depends on the event or incident at hand, ranging from preparing for a criminal case in court to identifying the root cause and impact of a cyber intrusion in a corporate network. We have adopted a definition based on Definition 1.1 on investigations above, encompassing both law enforcement and other applications.

Definition 1.5: Cyber investigations

A systematic collection, examination, and evaluation of all information relevant to establish the facts about an Internet-related incident or an alleged cybercrime and identify who may have committed or contributed to it.

1.3.1   Digital Forensics

A key component in any cyber investigation is digital forensics, which is the topic of our previous textbook (Årnes, 2018). In digital forensics, we process digital evidence (see below) according to well-defined scientific processes to establish facts that can help a court of law to conclude with regard to a criminal case. For the purpose of this book, we have adopted the definition by the National Institute of Standards and Technology (NIST) (2006).

Definition 1.6: Digital forensics

"The application of science to the identification, collection, examination, and analysis, of data while preserving the integrity of the information and maintaining a strict chain of custody for the data."

As a helpful reference, the Digital Forensics Workshop (DFRWS) proposed the more comprehensive definition in 2001: "*The use of scientifically derived and proven methods toward the preservation, collection, validation, identification, analysis, interpretation, documentation, and presentation of digital evidence derived from digital sources for the purpose of facilitating or furthering the reconstruction of events found to be criminal, or helping to anticipate unauthorized actions shown to be disruptive to planned operations.*"

1.3.2 Digital Evidence

The evidence addressed in digital forensics is referred to as digital evidence, "any digital data that contains reliable information that can support or refute a hypothesis of an incident or crime," as based on the definition originally proposed by Carrier and Spafford (2004).

Definition 1.7: Digital evidence
"Digital evidence is defined as any digital data that contains reliable information that can support or refute a hypothesis of an incident or crime."

In digital forensics, the two principles *chain of custody* (keeping a record of all actions when processing digital evidence) and *evidence integrity* (ensuring that evidence is not willfully or accidentally changed during the process) are central, as defined in Årnes (2018).

Definition 1.8: Chain of custody
Chain of Custody refers to the documentation of acquisition, control, analysis, analysis, and disposition of physical and electronic evidence.

Definition 1.9: Evidence integrity
Evidence integrity refers to the preservation of evidence in its original form.

1.3.3 Attribution

An essential question in cyber investigations is "who did it?", generally referred to as attribution, which is defined as "determining the identity or location of an attacker or an attacker's intermediary." The terms "traceback" or "source tracking" are frequently used terms addressing the same question (Larsen, 2003). However, it is generally understood that attribution is as much an art as a science (Buchanan, 2015), and advanced cyber-criminals will make efforts to hide their tracks in order to mislead investigators. Therefore, one should be very careful with concluding on an attribution without firm facts.

Definition 1.10: Attribution
"Determining the identity or location of an attacker or an attacker's intermediary." (Larsen, 2003).

1.3.4 Cyber Threat Intelligence

An important category of intelligence in cybercrime and cybersecurity is Cyber Threat Intelligence, as defined by the Center for Internet Security (2022). Cyber threat intelligence can help an investigator or forensic analyst identify and understand the evidence in a case in the context of information gathered from other sources.

Definition 1.11: Cyber threat intelligence

"Cyber threat intelligence is what cyber threat information becomes once it has been collected, evaluated in the context of its source and reliability, and analyzed through rigorous and structured tradecraft techniques by those with substantive expertise and access to all-source information."

1.3.5 Open-Source Intelligence (OSINT)

Another important category of intelligence is open-source intelligence (OSINT). We will adopt a slightly modified definition based on a recent publication by the RAND Corporation (Williams & Blum, 2018). A broad range of OSINT is available on the Internet and on the dark web, including technical information that is essential for attribution.

Definition 1.12: Open-source intelligence

We define OSINT as publicly available information discovered, determined to be of intelligence value, and disseminated by an intelligence function.

OSINT is further discussed in Section 4.5.

1.3.6 Operational Avalanche – A Real-World Example

In order to better understand the practical aspects of cyber investigations, this book will provide a variety of examples from known media and court cases. We start with "Operation Avalanche," a major international law enforcement operation resulting in the dismantling of the criminal infrastructure platform "Avalanche" in 2016 (Europol, 2016).

Example 1.1: Operation Avalanche

On November 30, 2016, after more than four years of investigation by FBI, Europol, Eurojust, several global partners, and prosecutors in the United States and Europe, the cybercrime platform Avalanche (Europol, 2016) was dismantled. Avalanche was used by cybercriminals worldwide for global malware attacks and money mule recruitment campaigns. The total monetary losses are estimated to be in the order of hundreds of millions of Euros.

(Continued)

Example 1.1: (Continued)

Criminal groups had used Avalanche since 2009 for malware, phishing, and spam activities. The criminal groups had sent more than 1 million infected emails to unsuspecting victims as part of their phishing campaigns. At any time, more than 500,000 computers were infected and part of the Avalanche network, with victims in more than 180 countries.

As detailed in Europol (2016), the investigation started in Germany in 2012 when encryption ransomware blocked users' access to their computer systems, and millions of computers were infected with malware enabling criminals to collect personal information (e.g., bank and email passwords). Avalanche enabled the criminals to perform bank transfers from victim accounts. A "double fast flux infrastructure" was employed to secure the proceeds and complicated the investigation.

Throughout the investigation, the German Federal Office for Information Security (BSI) and the Fraunhofer Institute (FKIE) collected and analyzed more than 130TB of captured data to identify the botnet infrastructure successfully. On the day of the Avalanche takedown, a command post at Europol was established, with representatives of the involved countries, Interpol, the Shadowserver Foundation, ICANN, domain registries, and several cybersecurity companies.

Throughout the investigation, prosecutors, and investigators from 30 countries were involved, resulting in 5 arrests, more than 30 premises searched, more than 30 servers seized, more than 200 servers taken offline, and more than 800,000 domains seized, "sinkholed" or blocked.

1.4 Challenges in Cyber Investigations

Cyber investigations remain a challenging and dynamic field of research, and practitioners are always racing to innovate investigative tools and methods in response to new threats, attack tools, and exploited vulnerabilities in an everlasting cat and mice game. In the research leading up to this book, the authors have identified several research questions that would benefit from additional research as an inspiration to students, with the objective of strengthening investigative capabilities:

- *Artificial intelligence (AI), machine learning (ML), and automation:* The use of AI and ML to automate the processing of large data volumes for investigative purposes for more effective cyber investigations and digital forensics. At the same time, CTAs are using AI and ML to automate cyber-attacks to improve their capabilities to reach their objectives and decrease the likelihood of successful detection, response, and investigations. How can we leverage AI, ML, and automation to increase the effectiveness of cyber investigations?
- *Internet of Things (IoT) and 5G*: IoT and 5G provide rapidly increasing numbers of connected devices with highly diverse use cases in complex ecosystems. While the technologies provide a new generation of security capabilities, they also represent a challenge for cyber investigations in terms of access to data and acquisition of digital evidence, ability to perform security monitoring and lawful intercept, as well as legal

and jurisdiction challenges. What are the challenges of 5G and IoT investigations, and how do we overcome them?

- *Operational coordination*: Cyber investigations are highly time-sensitive, and we discuss the importance of the golden hour of cyber investigations in Chapter 2. In order to succeed with cyber investigations, we are dependent on establishing efficient and, to a more significant degree, automated operational coordination for law enforcement, public-private cooperation, and cross-border data transfer for digital evidence. How can we enable cyber investigations through more efficient and automated operational coordination?
- *Attribution*: As we discuss in Section 1.3.3, attribution is a challenging, sometimes impossible process that requires careful consideration of the available evidence as part of the cyber investigations process, and there is an asymmetry between offensive and defensive capabilities (i.e., the threat actor has an advantage). Attribution, however, remains a critical objective for criminal, national intelligence, and incident investigations. Unfortunately, the uncertainty is often so high that the attribution is not openly disclosed and only stated as a hypothesis. How can we improve the confidence in attributions?
- *Standardization*: Cyber investigations and digital forensics depend on extensive data processing, but there is a lack of common standards for storage and exchange of digital evidence (Flaglien et al., 2011), and we are, to a large degree, dependent on proprietary systems. In order to enable automation and efficient operational coordination, improvements in standardization are required, ranging from forensic readiness standards (Dilijonaite, 2018) throughout the cyber investigation process. How can we adopt common standards for digital forensic readiness and cyber investigations?
- *Privacy vs. Security*: While privacy and security often have a common purpose (i.e., protecting data), there are also inherent conflicting objectives related to areas such as, on one hand, potentially privacy-intrusive technologies such as security monitoring, data loss prevention, and forensic readiness, and on the other hand privacy measures that hamper detection and investigations of crimes and incidents, such as encryption (this is often referred to as "going dark"), locked handsets and unavailability of who is registration data. How can we enable cyber investigations while maintaining both privacy and security?

1.5 Further Reading

We recommend that students of this book study supplementary literature to better understand Cyber Investigations. For this purpose, here is a list of relevant textbooks that are recommended by the editor and authors of this book:

- Årnes, A. (Ed.) (2018). *Digital Forensics*. John Wiley & Sons.
- Bazzel, M. (2018). *Open-Source Intelligence Techniques: Resources for Searching and Analyzing Online Information*. CreateSpace Independent Publishing Platform.
- Moore, M. (2016). *Cybersecurity Breaches and Issues Surrounding Online Threat Protection*. Information Science Reference.
- Bollinger, J., Enright, B., & Valites, M. (2015). *Crafting the InfoSec Playbook: Security Monitoring and Incident Response Plan*. O''Reilly.

- Golbeck, J. (2015). *Introduction to Social Media Investigation*. Syngress.
- Shipley, T., Bowker, A., & Selby, N. (2013). *Investigating Internet Crimes*. Syngress.
- Davidoff, S. & Ham, J. (2012). *Network Forensics: Tracking Hackers through Cyberspace*. Prentice-Hall.
- Clifford, R. D. (2011). *Cybercrime: The Investigation, Prosecution, and Defense of a Computer-Related Crime*. Carolina Academic Press.
- Stelfox, P. (2009). *Criminal Investigation: An Introduction to Principles and Practices*, 1st ed. Willan.
- Reyes, A., Brittson, R., O'Shea, K., & Steele, J. (2007). *Cyber Crime Investigations: Bridging the Gaps Between Security Professionals, Law Enforcement and Prosecutors*. Syngress.
- Stephenson, P. & Gilbert, K. (2004). *Investigating Computer Related Crime*, 2nd ed. CRC Press.

We have also provided extensive resources in the Educational Guide (Chapter 8).

1.6 Chapter Overview

This book is divided as follows:

- *Introduction*. Introducing the area of cyber investigations related to cyber security and cybercrime investigations. Clarifying the difference between forensics and investigations, both from a criminal and industry perspective. Introducing central definitions and modes of crime. (*Professor André Årnes, Ph.D.*)
- *Cyber Investigations Process*. Defining the investigation process step by step, mainly from a criminal investigation perspective. Discussing law enforcement cooperation across jurisdictions and current state-of-the-art research. (*Ph.D. student Nina Sunde, MSc*)
- *Cyber Investigation Law*. Legal aspects of cybersecurity and cybercrime investigations, following up on the Cybercrime Law chapter in the Digital Forensics book. Focus on the criminal process, evidence exchange, and cooperation between jurisdictions. Applications of data protection and national security law. (*Professor Inger Marie Sunde, Ph.D. and LL.M*)
- *Perspectives of Internet and Cryptocurrency Investigations*. The fundamentals of tracing and attribution on the Internet, building on the Internet Forensics chapter in the Digital Forensics book. Topics of interest include cloud computing, virtualization, cryptography, cryptocurrency, financial transactions, and open-source intelligence. (*Petter C. Bjelland, MSc*)
- *Anonymization Networks*. How do anonymization networks work, and how do they impact investigations? Addressing aspects of tracing, monitoring, evidence acquisition, de-anonymization, and large investigations. (*Associate Professor Lasse Øverlier, Ph.D.*)
- *IoT Investigations*. Performing investigations and digital forensics in the context of Internet of Things (IoT) technologies, including aspects of embedded systems, devices, connected systems, data fusion, and state-of-the-art research. (*Ph.D. Student Jens-Petter Sandvik, Cand. Scient.*)

- *Multimedia Forensics.* The role of multimedia (video, images, sound) in investigations, including how to leverage similarity matching, content-based tracing, and media metadata. (*Ph.D. Student Jens-Petter Sandvik, Cand. Scient. and Associate Professor Lasse Øverlier, Ph.D.*)
- *Educational Guide.* The Educational Guide includes guidance for teachers and students, with a wealth of references and practical information to be applied as part of research and studies. (*Kyle Porter, Ph.D.*)

1.7 Comments on Citation and Notation

For the benefit of the reader, the following standards have been adopted in the book:

- *Citations*: Citations to authoritative textbooks, research papers, and online sources are provided throughout. Students are encouraged to research the primary sources to understand the subject matter better.
- *Grey boxes*: Definitions, Examples, Legal Provisions, and Equations are highlighted in separate grey boxes. Examples can be either real-world case examples or illustrative scenarios.
- *Figures*: All photographs and illustrations are made by the chapter authors unless otherwise specified.
- *Software*: All software and hardware tools references are included as examples only. They do not represent a recommendation or preference regarding tool choices, and they should not be interpreted as guidelines or instructions on tool usage.
- *Index*: Key terms are indexed in the Index. All Definitions are included in the Index.

1.8 Exercises

1 Define Cyber Investigations and explain the difference between Cyber Investigations and Digital Forensics. Provide examples to illustrate the two concepts.

2 What is Cybercrime? Explain the two main interpretations of cybercrime.

3 What is a Cyber Threat Actor? Define the term and detail the simplified three-tier model used in this book.

4 What is the Cyber Kill Chain? Explain why it is essential and explain and provide examples of its seven phases.

5 What is MITRE ATT&CK™? Provide a definition and explain the main elements of the framework. How can MITRE ATT&CK™ be applied in a digital investigation?

6 What are the two main principles of Digital Forensics, and why are they essential to Cyber Investigations?

7 What is attribution, and how can it be addressed during an investigation? Explain why attribution is challenging and propose steps to increase the level of confidence in the attribution in an investigation.

8 Explain how intelligence can support Cyber Investigations and provide examples of Threat Intelligence and Open-Source Intelligence sources.

9 Summarize "Operation Avalanche" and outline the most critical challenges law enforcement must address during such a complex and long-lasting investigation.

2

Cyber Investigation Process

Nina Sunde*

Department for Post Graduate Education and Training, The Norwegian Police University College, Oslo, Norway

Conducting a high-quality and effective cyber investigation can be a complex task. Different competencies are necessary to uncover and produce relevant and credible evidence about what has happened, by whom, against whom, and with which purpose. At least two processes are involved in investigating cyber-dependent or cyber-enabled crimes – the criminal investigation and the digital forensic processes. These processes originate from different scholarly traditions and are often described in isolation. This chapter presents a novel concept that integrates the two processes: the *integrated cyber investigation process* (ICIP) aimed to facilitate structured and well-coordinated cyber investigations.

The procedural stages of ICIP dynamically interact with the concept we have named *The Cyber Investigation Queries*, a model for addressing the different phenomena of cybercrime and guides the systematic development and testing of investigative hypotheses. We emphasize that the human factor is essential for reaching the goal of a high-quality and fair investigation. At the same time, we recognize the need for awareness and domain-adequate measures to minimize the risk of human error during cyber investigations.

2.1 Introduction

The Internet was initially created for military and scientific purposes, and nonexperts did not widely use it until the early 1990s (Choi *et al.*, 2020). Expanded bandwidth, increased reliability, reduced surfing fees, and user-friendly interfaces are factors that have made cyberspace more accessible to the public (Curran, 2016; Greenstein, 2015). However, technological changes can result in new types of crimes, and as people move – crime follows. As Internet access has become more widespread, crime in digital spaces has begun to affect individuals and organizations in new ways adversely.

---

* Nina Sunde is a Police Superintendent and Lecturer at the Norwegian Police University College and a Ph.D. student in Criminology at the Department of Criminology and Sociology of Law at the University of Oslo. Nina holds an MSc in Digital Forensics and Cybercrime Investigation from NTNU.

Cyber Investigations: A Research Based Introduction for Advanced Studies, First Edition. Edited by André Årnes.
© 2023 John Wiley & Sons Ltd. Published 2023 by John Wiley & Sons Ltd.

However, crime associated with technology did not "begin" with the Internet. For example, the factories and railroads in the first industrial revolution enabled train robberies. Along with the electricity and cars from the second industrial revolution, car thefts emerged (Choi *et al.*, 2020). The third industrial revolution is linked to an interconnected society enabled by computers and the Internet, with crimes such as hacking, cyber theft, and malware development and distribution (Choi *et al.*, 2020). The fourth Industrial Revolution is currently being shaped and developed using the Internet and technology such as Internet of things (IoT), cryptocurrency, and artificial intelligence (Schwab, 2016). With these technologies, new forms of crime appeared, such as fraud by using deep fake technology and cryptocurrency ransomware (Choi *et al.*, 2020).

Alongside the technology development and the rise of novel crime phenomena, law enforcement has changed to control and investigate it. Technology has provided new tools and techniques for the police, such as surveillance, analysis of big data, and crime prediction. Forensic science has evolved, with advancements in, e.g., DNA technology. Investigating cyber-enabled and cyber-dependent crimes have necessitated the police to develop in-house technological expertise or hire experts to investigate such crimes. The police must continuously develop and update their knowledge about the crime phenomenon and modus operandi to effectively investigate these novel crime types. Such knowledge is crucial for predicting where relevant evidence may be located and the typical perpetrators or victims.

For a long time, a criminal investigation was perceived as a craft learned through experience and by observing more experienced colleagues, or an art – where instincts, intuition, and feelings played an important role (Hald & Rønn, 2013). However, the field has matured and is today a profession with a more research-based foundation derived from or inspired by other areas than what may be perceived as "pure" policing, such as the scientific methodology, sociology, criminology, psychology, philosophy, and technology (Bjerknes & Fahsing, 2018).

An effective cyber investigation requires many different knowledge components, mainly from criminal investigation methodology, law, and technology (Sunde, 2017). However, merely mixing these components will not necessarily lead to success. High-quality and effective cyber investigations require the right people are at the right place, at the right time, and do the right things correctly and for the right reasons. An adequate framework for collaboration and coordination is necessary to achieve this goal.

This chapter presents a novel concept for cyber investigation: the ICIP, which includes the following stages: Investigation initiation, modeling, planning and prioritization, impact and risk assessment, action and collection, analysis and integration, documentation, and presentation, and evaluation. The procedural stages of ICIP are in dynamic interaction with the model that we have named The Cyber Investigation Queries, where the different phenomenological components of cybercrime are addressed, facilitating the systematic development, and testing of investigative hypotheses.

The following section will zoom in on what investigation is, and Section 2.3 presents and discusses the models that have inspired the development of ICIP. The principles on which ICIP is based are explained in Section 2.4, and its procedural stages are presented and exemplified in Section 2.5. Section 2.6 describes cognitive and human factors relevant to a cyber investigation. We elaborate upon how these factors may be sources of bias and appropriate measures to minimize bias.

2.2   Investigation as Information Work

The criminologist Martin Innes states that criminal investigations, in essence, are *information work,* acted out in a distinct order, with the primary function of reducing uncertainty (see Innes, 2003, 2007). Innes (2003) described three interconnected movements that are present in the order of an investigation:

1) Identify and acquire
2) Interpret and understand
3) Order and represent

A criminal investigation often starts with bits and pieces of uncertain, unverified, and incomplete information. The threshold for when the information can be acted upon depends on the context and purpose. Within a criminal investigation, the systematic work toward reducing uncertainty is closely connected to the legal evidential requirement for conviction, often expressed by the phrase "proven beyond any reasonable doubt." Within other contexts (military, private industry), the primary function of the investigation order would often be the same. However, the thresholds for probability or certainty before the information is acted upon may be different.

Within an investigation, the traces are converted into several *key modes* (Maguire, 2003). Data ordered and communicated can be defined as *information* (Innes, 2003). When the relevance and credibility of the information are established, the information develops a factual status of *knowledge*. Information of varying provenance that can be used internally by the police organization to plan future actions and lines of inquiry is *intelligence. Evidence* is information assembled into a format suitable for use in the legal process.

We will now move from the high-level abstraction model of investigation to a more detailed framework for cyber investigation.

2.3   Developing an Integrated Framework for Cyber Investigations

Conducting a cyber investigation will involve at least two parallel processes: first, the investigation process (or order) (see Fahsing, 2016; Innes, 2003, 2007) concerned with the "tactical" investigation of various evidence types, such as testimonial and tangible evidence and second, the digital forensic process (Flaglien, 2018) where digital evidence is handled. These processes should be conducted by personnel with specialized compe tence and experience. To ensure close cooperation between the personnel involved in the two processes, a mutual understanding of the common goal and a well-defined structure for cooperation are necessary. The following section will describe an integrated framework for cyber investigation, where the investigation process and the digital forensic process are assembled in a joined structure. The framework is named the *ICIP*. Before describing its procedural stages, the rationale behind ICIP and the principles on which it is built are outlined.

One of the first to move from a narrow focus on the processing of digital evidence to cybercrime investigation was (Ciardhuáin, 2004) with the "Extended model of cybercrime investigation," which had particular attention toward the information flow within the investigation. Later, Hunton (2009, 2011a, 2011b) developed the "Cybercrime investigation framework," which aimed to bridge the gap between technology examination and law enforcement investigation. The ICIP is a further development toward integrating the processes of criminal investigation and technology examination, building on elements from the models developed by Hunton, Innes (2003, 2007), and Fahsing (2016).

Hunton's Cybercrime Investigation Framework was mainly based on principles from several Association of Chief Police Officers (ACPO) guidelines, such as ACPO core investigative doctrine (ACPO, 2005). One of these principles is to obtain an *investigative mindset*, which involves conducting a structured investigation through the steps of first – understanding the source, second – planning and preparing, third – examining, fourth – recording and collating, and eventually – evaluating (Jones *et al.*, 2008). An investigative mindset is associated with the assume nothing, believe nothing, challenge everything (ABC) rule and aims to safeguard a major incident's successful and ethical investigation.

Definition 2.1: The ABC rule (Cook, 2016)
• Assume nothing • Believe nothing • Challenge/check everything

However, the ACPO core investigative doctrine has been criticized for not being evidence-based and not providing detailed guidance on thinking and deciding during a criminal investigation (Fahsing, 2016). Fahsing points out the many pitfalls caused by cognitive heuristics and biases and suggests The Investigative Cycle as a model for diagnostic decision-making. The Investigative Cycle is a further development of Innes' description of the steps in major inquiries in England and Wales (Innes, 2003) and the 5c's (collect, check, consider, connect, construct) described by Dean (2000) after studying Australian detectives. In addition, Fahsing added the sixth step – consult – and remodeled it into a cyclic process (Figure 2.1).

Although The Investigative Cycle aims to prevent bias, one of the model's limitations is that it does not acknowledge the investigator's subconscious emerging hypothesis of what has happened, which will affect the decisions being made from the beginning of an investigation (see Sections 2.5.1 and 2.6). The forming and testing of hypotheses start in stage 4 (construct), when extensive information has already been collected and evaluated (checked). The ICIP addresses this issue with a structured approach from the very beginning to reduce confirmation bias. The initial hypothesis (e.g. the system is hacked, and information is stolen) should be acknowledged

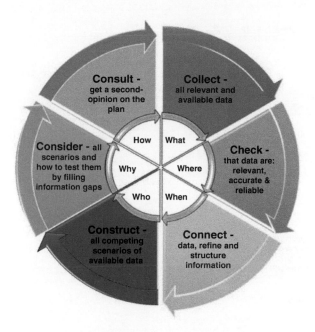

Figure 2.1 The Investigative Cycle (Fahsing, 2016).

and documented. The investigator should also define and document the opposite or negation of the hypothesis (the system is not hacked). This strategy ensures that at least two competing hypotheses form the basis for the investigation from the very start.

A strength of The Investigative Cycle is the focus on obtaining a second opinion. Consulting someone from outside the investigation team is essential for preventing tunnel vision. By introducing this as a measure that is repeated several times during the investigation, it allows for the possibility to detect and correct errors before they impact the result of the investigation.

Evaluation can be performed top-down or implicit as a deep sub-process within other core processes (Pollitt *et al.*, 2018). The proposed ICIP framework acknowledges the need for informal evaluation and second opinion consultations during the investigation. Therefore, each stage of ICIP involves the encouragement of consulting someone independent of the investigation. At the same time, a more formalized evaluation stage is included as the final stage to facilitate learning from both the accomplishments and the failures of the investigation.

Active use of the six basic queries who, what, when, why, where, and how, often themed the 5WH (see Section 1.1), can be a helpful strategy for establishing the case circumstances. These queries are related to a cybercrime context in the Cyber Investigation Queries. The model aims to serve as a effective thinking aid for developing plausible hypotheses based on the available information. The hypotheses guide the further investigation and are used for identifying information suitable for testing the hypotheses, and predicting where the information may be located. This model should therefore be actively used during all the stages of ICIP (Figure 2.2).

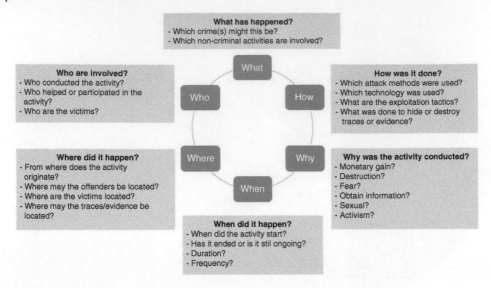

Figure 2.2 The cyber investigation queries.

2.4 Principles for the Integrated Cyber Investigation Process (ICIP)

The ICIP draws on theories and principles from several scholarly traditions, such as criminal investigation, forensic science, digital forensics, forensic psychology, scientific methodology, and law. The core principles and theoretical components of the ICIP framework are described below.

2.4.1 Procedure and Policy

All investigative activities should be conducted in compliance with human rights, with particular attention to the right to a fair trial (European Court of Human Rights (ECHR), article 6 no. 1) and to be presumed innocent (ECHR, article 6 no. 2). All actions should be done in compliance with the criminal procedural rules within the jurisdiction the investigation is carried out and applicable ethical guidelines.

2.4.2 Planning and Documentation

An investigation plan and a decision log should be created and continuously updated in every investigation. The time spent detailing an investigation plan and decision log is likely a good investment in investigative quality and efficiency. The scope and purpose of each task and action should be described in detail to avoid misinterpretations and misconceptions. By explicitly assigning tasks to personnel, the responsibility of each individual involved in the investigation is made visible.

In addition to scaffolding the ongoing investigation, the documentation also serves a second purpose, namely evaluation. The investigation plan and decision log facilitate review from third parties during the investigation, which may provide an opportunity to correct a course heading in the wrong direction. After the investigation is completed, these documents constitute a valuable source for organizational development and learning based on knowledge derived from the investigation's mistakes as well as its successes.

2.4.3 Forming and Testing of Hypotheses

It is vital, to avoid flaws in the investigation, that personnel with adequate competence (investigative/technological/law) conduct the investigative assessments, decisions, or actions. Of equal importance to the competence is the investigative mindset, and the investigators should therefore do their best to comply with the ABC rule (see Definition 2.1).

The forming and testing of hypotheses throughout the investigation is a fundamental part of the ICIP framework. In a criminal investigation, the hypotheses may be defined on different levels, from the general and overarching offense level to the narrower activity and source levels. This way of structuring investigative hypotheses originates from the forensic science domain and is primarily related to evaluative reporting of expert opinions (Cook *et al.*, 1998; ENFSI, 2015a; Jackson *et al.*, 2015).

- *The offense level* is the most general level, and the hypotheses relate to the incident or criminal offense under investigation. The offense level hypotheses may include "objective" elements, such as what crime may have happened, and the "subjective" elements concerning criminal intent and guilt (Jackson, 2011). An example of a hypothesis centered on the objective elements of the crime is "*someone hacked the network of firm X and stole confidential information.*" A hypothesis concerned with criminal guilt may be articulated as "*the suspect was aware that the content she intentionally downloaded from the server was illegal.*" The offense level hypotheses are also referred to as investigative hypotheses further in the chapter.
- *The activity level* hypotheses are centered on the performed activities, independent of whether they constitute criminal activities. For example, "*social engineering techniques were used to gain access to the firm X's network,*" or "*the malware program was sent from the email address xxx@yyy.com to the mail address yyy@zzz.com.*"
- *The source-level* hypotheses relate to the source of identified traces. In forensic science, it is sometimes necessary to include an additional sub-source level. For example, "*the suspect X used the email address xxx@yyy.com when sending the malware*" or "*the user account X performed the malware downloading.*"

An exhaustive set of investigative hypotheses at the offense level should be defined in the investigation plan and be continuously updated. They form the basis for defining and testing hypotheses at lower levels. Therefore, it might be necessary to define sub-hypotheses at the activity and source levels to test an offense level hypothesis adequately. For example, to prove that the suspect hacked the network of firm X and stole information (offense level), the investigation must uncover the activities performed by the suspect which constituted the crime (activity), and the evidence must be linked to the suspect and firm X (source).

To operationalize the presumption of innocence (see, for example Stumer, 2010), the set of hypotheses should, as a minimum, include plausible noncriminal scenarios explaining the initial information. For example, *"The firm X was not hacked, and no information was stolen."* Even if a suspect is pointed out, the set should include a hypothesis about his or her innocence as a minimum. For example, *"The suspect has nothing to do with the hacking and information theft at firm X."*

Several approaches for hypothesis testing exist, and here we will include the *Bayesian* and *falsification* approaches. Very simplified, the *Bayesian approach* aims to compute or establish likelihoods, while the *falsification approach* eliminates hypotheses through thorough falsification attempts.

The Bayesian approach aims to determine evidential strength relative to the hypotheses and case circumstances (Aitken *et al.*, 2010). Sometimes a likelihood ratio is calculated, which should always be expressed as "strength of evidence relative to the hypotheses" approach, as opposed to the potentially misleading "strength of hypothesis" approach (Casey, 2020; ENFSI, 2015a). The strength of hypotheses given the evidence approach is also known as "the prosecutor's fallacy" (Thompson & Schumann, 1987), which states that an investigator may be misled to believe that the hypothesis is highly probable. For example, the probability that someone knows how to script given that he or she is a hacker does not necessarily equal the probability that someone is a hacker given that he or she can script. For more examples – see Section 2.5.7.

Through the falsification approach, the hypotheses are eliminated through actively seeking information that contradicts or disproves the hypothesis (Popper, 1963). In a criminal investigation, falsification in a strict sense is often not achievable due to the nature of the available information. However, the threshold for proving guilt in the criminal justice system is different from establishing "the truth" in science. The threshold for guilt in the criminal justice system is often expressed as "beyond reasonable doubt."

2.4.4 The Dynamics of ICIP

The dynamics of the ICIP make it a flexible process. It is cyclical, meaning that the process may be repeated until the goal of the investigation is achieved. ICIP is also nonlinear, making it possible to move back to the previous stages, if necessary, as indicated by the arrows between the stages (see Figure 2.3). This implies that all the stages may be revisited several times. However, this flexibility should not be understood as an opportunity to skip any stages, such as moving directly from initial investigation to action. The forming and testing of investigative hypotheses will require that the investigation team use the Cyber Investigation Queries several times until all the investigative hypotheses are sufficiently tested.

After comprehensive information collection and hypothesis testing, the legal decision-makers evaluate the result against the evidentiary requirement "proven beyond reasonable doubt." To substantiate criminal guilt beyond reasonable doubt, the (relevant and credible) evidence should be consistent with a plausible criminal scenario, and the evidence should *not* be consistent with a plausible innocence scenario.

2.4.5 Principles for Handling Digital Evidence

When handling digital evidence, obtaining and preserving evidence integrity (see Chapter 1, Definition 1.9) should always be the leading principle. Deviation from this principle must be justified and documented, and the consequences should be explained as detailed as possible. All evidence, including digital evidence, should be handled in

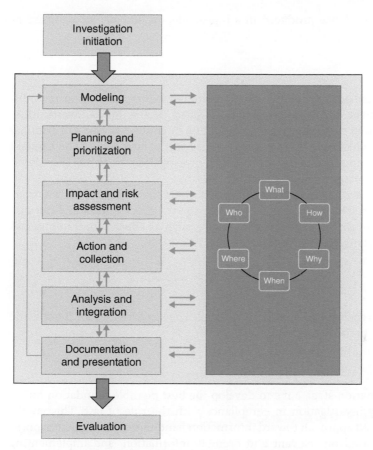

Figure 2.3 The Integrated Cyber Investigation Process (ICIP).

compliance with the principle of an audit trail, and documentation of the chain of custody (see Chapter 1, Definition 1.8: Chain of Custody) should always be kept in an updated state.

Digital evidence is transformed from digital traces to meaningful evidence in a legal context by the people involved in the investigation. Therefore, such evidence is prone to both human and technical error during this process, and error mitigation measures should be introduced to prevent or detect flawed evidence. The Scientific Working Group on Digital Evidence (SWGDE) guideline *Establishing Confidence in Digital Forensic Results by Error Mitigation Analysis* provides helpful guidance for minimizing technical error (SWGDE, 2018).

For mitigation of human error, see Section 2.6. The investigators should be transparent about any factors that may have impacted the *evidential value*, which consists of two key components: *relevance* and *credibility*. Evidential relevance means that the evidence is suitable to make one or more of the investigative hypotheses probable or sheds light on the credibility of the evidence (Anderson *et al.*, 2005). Three attributes must be assessed to establish evidential credibility (Anderson *et al.*, 2005):

- *Authenticity* – is the evidence what it purports to be?
- *Accuracy/sensitivity* – does the evidence provide the necessary resolution or completeness required to discriminate between possible events/explanations?

– *Reliability* – is the evidence produced in a repeatable, dependable, and consistent process?

2.4.6 Limitations

Although this book is aimed at cyber investigations conducted by private industry, military, or law enforcement, the ICIP is primarily related to a criminal investigation. However, evidence seized through private investigation might be relevant to a criminal investigation. For the evidence to be admissible in the court of law, it should be handled according to the rules and regulations for a criminal investigation. Note that the admissibility rules vary between different jurisdictions.

The ICIP framework scaffolds cyber investigations within the organizational environment it is implemented. This chapter will only describe the process itself and not discuss the implications of the organizational structure, such as education, training, and establishing policies, plans, and procedures (Figure 2.3).

2.5 ICIP's Procedural Stages

This section describes each stage of the ICIP in detail. The key activities for the specific stage are summarized in the definition boxes.

2.5.1 Investigation Initiation

The Investigation initiation stage aims to develop the best possible foundation for an effective, high-quality investigation in compliance with the rule of law. This entails establishing a structured approach toward information handling, ensuring that hypotheses are developed based on relevant and credible information, and implementing measures to prevent cognitive and human error from the very beginning.

Definition 2.2: Investigation initiation stage

The key tasks of the investigation initiation stage are:

- Get an overview of the available information.
- Assess whether this information is relevant and credible.
- Identify the initial investigative hypothesis at the offense level (what has happened) and the competing hypothesis. Define both in the investigation plan.
- Plan and conduct the initial actions to collect relevant information and verify the available information.

Knowledge of a crime might be a result of a report from the public or criminal intelligence. Regardless of what initiated the investigation, the investigator must obtain oversight and control over the information to establish what is known and identify any information gaps. Relevant here is, for example, to establish what initiated the report, what actions have already been taken, what source material and technical data are potentially available, and the technical elements of the potential offense (Hunton, 2011a, 2011b). Every new piece of information should be assessed by asking: *Is the information relevant and credible?* Considering the information pieces in context provides insight into the information's completeness and establishes what is known and what is missing or uncertain.

Despite the attempt to be objective and to obtain an "investigative mindset," the first hypothesis of what has happened will often evolve subconsciously. This may lead to a confirmation bias (see Section 2.6), and solely relying on awareness to counter this powerful bias is an inadequate measure (Dror, 2020). Therefore, the investigator will need a strategy to counter the confirmation bias that might affect the observations and decisions throughout the investigation (see Section 2.6.3). The essence of such a strategy is that it should constantly remind the investigator to consider alternative hypotheses throughout the investigation and commit him/her to evaluate each piece of evidence against all the hypotheses. The only valid reason for drawing attention away from one or more of the hypotheses at this stage should be if they have been refuted or falsified (see Definition 2.3). When the first hypothesis has emerged in the investigator's mind, it should be written down, together with at least one contradictory hypothesis established by the approach: what if Hypothesis 1 is not true – what may the scenario be, or simply the negation of Hypothesis 1.

Definition 2.3:   Refuting a hypothesis
Refuting a hypothesis means to falsify or disprove it. Refuting is an effective approach since we could collect a massive amount of information that supports the hypothesis, but we would only need one piece of (credible) information to disprove it.

Benjamin Franklin said, "By failing to prepare, you are preparing to fail," and later, Winston Churchill modified the phrase to "He who fails to plan is planning to fail." This proposition is also valid for criminal investigation, regardless of its scope. The investigation plan could be a plain text document, a spreadsheet, or a database. However, the critical point is whether the chosen tool fits the scope of the investigation. A lengthy and large-scale investigation would need a tool with more functionality, such as filtering, sorting, and searching functionalities, compared to a routine, small-scaled investigation. The main components of an investigation plan would be:

- The investigative hypotheses (and the paragraphs in the penal code to which they relate)
- The information requirement (what information is needed to test the hypotheses)
- A formulation of the tasks to fulfill the information requirement
  - The purpose of the task (which information requirement they refer to)
  - Identification of where the information might be located
  - The activities/methods/tools that may be used to collect the information (including coercive measures)
- The person responsible for undertaking the task
- Deadlines for the investigation and the defined activities/tasks

A decision log should also be established as an aid for decision-making and recollection (Stelfox, 2013). Not all investigative decisions are stated in documents and filed in the case, and a decision log will help the investigator to remember why specific strategies were chosen and on which grounds certain decisions were based. The investigation plan and the decision log also play a vital role at the evaluation stage (see Section 2.5.8).

"The golden hour" (Innes, 2007; Stelfox, 2013) is regarded as critical for a successful investigation and adheres to the need for immediate action to secure and preserve evidence that otherwise would be lost (see Definition 2.4). The golden hour is particularly important for cyber investigations since digital traces may be transient with a high risk of alteration or deletion. Hence, the sources that may carry relevant information must be identified and seized as soon as possible. However, the need for speed must not compromise the obligation to handle the evidence in a manner that preserves the evidential value by preserving the evidence integrity (if possible), maintaining a chain of custody, and ensuring transparency in all the investigative actions.

Definition 2.4: The golden hour

The golden hour is the period immediately after a crime was committed when material is readily available in high volumes to the police (College of Policing, 2020). Prioritizing evidence collection during the golden hour will maximize the quantity of relevant material gathered and the quality of the information since it is secured before it is substantially altered or lost. For example, a witness may have a clear memory of an event right after it happened. However, if the police wait too long, the witness may have forgotten details of the event, and the memory of the event may have changed due to the influence of internal or external factors.

Example 2.1: The Call4U case – investigation initiation stage

Case example – Call4U
Please note: This is a fabricated case example for illustration of the procedural stages of ICIP.

Investigation initiation:
The Chief Technology Officer (CTO) at the Telecom company Call4U contacts The National Cybercrime Centre (NC3). The CTO is concerned that they have been hacked and that confidential information has been stolen. The background for the concern is that several employees reported that they had received emails with attachments containing what is believed to be malware. Several of the employees had opened the attachments. The technology department had noticed unusual activities during the last days when inspecting the company's logs. The logs and suspicious emails are provided to the NC3, which decides to start a criminal investigation of the incident.

The investigation team members are carefully selected due to the competency needed for a high-quality and effective investigation. The team possesses knowledge and experience from cybercrime investigations, digital/network forensics, and law. They assess the information and decide that the best possible guess is that this may be a spear-phishing attack based on the current information. They write down the hypothesis and consider the alternative hypothesis could be if the first one was not true. They agree that the incident also may result from several coincidences and that the suspicious emails and the unusual log activities may be unrelated.

Hypotheses:

Spear phishing attack vs. No relation between suspicious emails and log activity
The team decides that the following actions need to be conducted to check the initial information more thoroughly:

- Task 1: Get an overview of the suspicious email activity (time period, who received them)
- Task 2: Examine the email attachment. Determine if malware really exists in the attachment and what it does
- Task 3: Examine the log files
- Task 4: Interview the CTO
- Task 5: Interview the employee who discovered the unusual activity

2.5.2 Modeling

The modeling stage aims to develop a broad set of investigative hypotheses at the offense level, to guide the identification of information requirements and prevent tunnel vision.

Definition 2.5: Modeling stage

The key tasks of the modeling stage are:

- Link new information to the existing and identify information gaps.
- Define all relevant investigative hypotheses at the offense level by using The Cyber Investigation Queries.
- Seek a second opinion on whether all relevant hypotheses have been identified.

Modeling relates to the second motion within the investigation order (Innes, 2007), where the information is interpreted into intelligence or knowledge (see Section 2.2). First, an updated overview of the available information should be obtained, and the pieces of information should be linked together, for example, in a timeline. This will provide an overview of what is known and insight into the information gaps. The set of hypotheses defined in the previous stage should now be expanded by working through the Cyber Investigation Queries. All relevant investigative hypotheses at the offense level should be developed, and the result should cover a spectrum from the worst case to innocence hypotheses.

Due to the reliance on contingent and uncertain information, the sense-making and forming of hypotheses will be by *abductive inference*, which together with deductive and inductive inference represent the three methods of logic inference. Therefore, the formation of a hypothesis will be *reasoning to the best explanation* (Josephson &

Josephson, 1996), which may be done by answering the question: *What are the best and most plausible reasons, given what is currently known, to explain how this information and circumstances came to be?* The hypotheses will be formed based on the existing case information, the investigators' domain-specific knowledge, as well as their general world knowledge (Kerstholt & Eikelboom, 2007; Sunde, 2020).

A particular risk at this stage is *guilt bias* (see Section 2.6), which entails more attention toward hypotheses involving criminality and guilt and less attention to hypotheses involving innocence or mitigating circumstances. To minimize the guilt bias and safeguard the presumption of innocence, the investigation team should always ensure that one or more noncriminal or innocence hypotheses are present in the set of hypotheses.

The modeling stage is not a static nor an isolated stage of an investigation. The modeling stage should be revisited when new information is collected to update the hypotheses. The emerging narratives (e.g. the timeline) should be updated and extended by linking new information to what is already known (Innes, 2007).

Example 2.2: The Call4U case – modeling stage

Case example – Call4U

Modelling stage:
Based on the information obtained during the Investigation initiation stage, the investigation team at NC3 has a more detailed picture of what may have happened. The credibility of some of the information is checked. What they know by now may be summarized to:

- According to the CTO, Call4U was about to launch a new business strategy directed toward a new group of customers. The CTO feared the attack was aimed to obtain the details of the business strategy and the identity of the potential customers.
- An employee at the tech department was notified about unusual activities in the corporate network by a system alarm. He examined the web logs and noticed that six employees received an email from the same Internet Protocol (IP) address the night before the unusual activity started. He did not know if the same activity had happened before and informed the CTO that they would monitor the system for future similar attacks.
- The malicious email attachment was a Portable Document Format (PDF) file. A preliminary examination of the source code indicated that opening the PDF would execute a code that would assign administrator rights and allow access to restricted areas of the system.
- The log files indicated that information was sent to several external email addresses during unusual times, but the CTO indicated that it would take some time to obtain an overview of this activity.

The investigation team uses the Cyber Investigation Queries to expand the set of hypotheses due to the current information basis. When the team is done, they consult a colleague from another team. Based on the information and her own experience from a similar case, she suggested an additional hypothesis. The result of the hypotheses development is:

Hypotheses:
H1 Spear phishing attack against Call4U

- H1.1 aimed at gaining information
- H1.2 aimed at harming the company (destroying reputation)

H2 Organized spear-phishing attack – more companies are attacked
H3 An insider from Call4U has conducted or contributed to the crime
H4 No relation between suspicious mails and log activity
H5 No criminal activity has occurred
- H4.1 The mail attachments did not contain malicious code
- H4.2 The suspicious log activity did not result from criminal activity

2.5.3 Planning and Prioritization

The planning and prioritization stage aims to define a clear and concise information requirement suitable for testing the hypotheses. The information requirement should serve as guidance to define where the information may be found and which methods or actions could be implemented to seize the information in an effective and high-quality manner.

Definition 2.6: Planning and prioritization stage

The key tasks of the planning and prioritization stage are:

- Define information requirements for testing the hypotheses.
- Identify where the sources of the relevant information may be located.
- List actions and methods that may be used to obtain relevant information.
- Plan and prioritize the investigative and technical tasks.
- If a suspect is identified based on an IP address: assess who else could have used the IP address prior to entering the search scene.
- Assess any legal or policy issues related to the planned actions and methods.
- Assess whether the necessary resources, competence, tools, and technology are available.
- Seek a second opinion on the plan and the prioritizations.

The result from the modeling stage should be a set of hypotheses in the written investigation plan. It is now time to define the information requirements, which may be identified by considering each of the hypotheses based on the following question: *If this hypothesis was true, which information must be- or would usually be present?* Now the possible sources or sensors should be identified by considering each piece of information and asking: *Where may this information be found?*

From an efficiency point of view, obtaining information with diagnostic value should have high priority. The diagnostic value depends on the information's usefulness for supporting or refuting the hypotheses. Information that is consistent with one hypothesis and contradicts the other has diagnostic value, in contrast to information consistent with all the hypotheses (Heuer, 1999).

Before any action is carried out, careful planning should be conducted. The investigation management must consider whether they have the necessary time, personnel, tools, and technology to perform the planned actions. The management must also ensure that the personnel has the right competence to undertake the planned actions and sufficient expertise to use the tools and technology to collect the evidence in a forensically sound manner. It is also important to consider any unsolved legal, ethical or policy issues with the planned actions.

A recurring challenge in cyber investigations is attribution (see Definition 1.10), which entails determining *who* committed a crime. An IP address only identifies a location, and a Media Access Protocol (MAC) address only identifies a networked device. This information does not identify the person using the device or the network, and this is very important to bear in mind when entering a search scene. Sometimes, information on the search scene makes it necessary to reconsider the hypotheses about who has committed the crime.

All actions must comply with the law, which for cyber investigations might sometimes be a complicated issue. The location of a computer and the data available from it do not necessarily correspond, and networked technology does not indicate when borders are crossed. The investigator must, therefore, pay close attention and be very careful to avoid conducting an investigation outside own jurisdiction. When such investigative measures are necessary, international cooperation with other law enforcement agencies should be initiated based on relevant legal instruments.

Both the investigative and technical tasks should be planned. Some investigative considerations are: If the suspect is present on the search scene, what information should he or she receive? Should the suspect be interviewed before arrest? If people unrelated to the case are present, how should they be handled, for example, in terms of identification, search, or interview? How should access to the system/network be obtained? What types of information (other than data) is of interest, for example, documents, situational tracks, and physical traces? Should other forensic evidence be secured, such as fingerprints, and DNA? Should the search scene be filmed or photographed prior to the search?

Some technical considerations are: What items are expected, and how should they be seized to maintain evidence integrity? Should an examination be undertaken on the search scene? If a device or system is running, how should it be seized to preserve the traces of interest? How should the search be documented to safeguard transparency and chain of custody?

Prioritizing is necessary to ensure prompt identification and collection of the most critical information. If the operation involves a risk to a victim's life or health, or to avert a serious crime, obtaining information for locating the scene and the people involved must be prioritized. In other situations, the prioritization should be concerned with the following questions:

- Which information must be seized immediately that might otherwise be lost or lose its evidential value?
- Which information has the highest diagnostic value?
- Which information may fill the most critical information gaps?

The investigation plan and decision log should be updated with information about the planned actions and decisions. When the planning and prioritization are undertaken, a second opinion from someone outside the investigation team should be obtained.

Example 2.3: The Call4U case – planning and prioritization stage

Case example – Call4U

Planning and prioritization stage:

Information requirement:	Current information/knowledge:
What has happened?	H1–H2–H3–H4
How was it done?	Spear phishing, other methods?
Which technology was used?	Malware email attachment
What are the exploitation tactics?	Escalation of rights gave access to the system
What was done to hide/destroy traces/evidence?	Unknown
When did it happen?	
When did the activity start?	Two days ago
Are there traces of former attempts?	Unknown
Has it ended, or is it still ongoing?	Unknown, is being monitored
Duration – frequency?	Unknown
Why was the activity conducted?	Get a business advantage? Harm?
Where did it happen?	
From where does the activity originate?	IP xxx.xxx.xxx.xxx of the emails with malware
Where may the offenders be located?	Unknown
Where are the victims located?	Call4U, other companies?
Where may the trace evidence be located?	At Call4U, at the offender's computer? The external email accounts to which the information was sent out?
Who is involved?	
Who conducted the activity?	Unknown
Who helped/participated in the activity?	Unknown
Who are the victims?	Call4U, other companies?

Actions/methods and prioritization:

The following actions/methods are planned to close the information gaps or control the information's credibility:

– Task 6: Trace the suspicious incoming emails containing malware. Priority: high.
 Task 7. Examine and test the malware to determine what it may have done/facilitated. Priority: high.

(Continued)

Example 2.3: (Continued)

- Task 8: Trace the outgoing emails to determine who they were sent to and which information they contained. Priority: high.
- Task 9: Collect, secure, and analyze information from the computers that received the malware. Look for traces of activities related to the suspected attack and traces of the malware. Priority: high.
- Task 10: Check if more companies have been attacked with the same malware. Priority: medium.
- Task 11: Establish whether the same activity has occurred before or is still ongoing. Priority: high.

A colleague from a different investigation team is consulted, and after a discussion, the priority of task 10 is changed from high to medium.

2.5.4 Impact and Risk Assessment

The impact and risk assessment stage aims to minimize all potential and known risks that could harm the parties involved in the investigation or affect the evidence's value.

Definition 2.7: Impact and risk assessment stage

The key tasks of the impact and risk assessment stage are:

- Consider legislation and policy implications for the planned actions.
- Consider the ethical aspects of the planned actions.
- Assess the technical skills of the suspect.
- Assess the risk of violence or other types of harm.
- Seek a second opinion about whether you have identified and mitigated all the possible risks and impacts for the planned actions.

Risk is the effect of uncertainty on objectives and is described as the consequence of an event and the associated likelihood of occurrence. The investigation team should identify risks and carry out necessary measures to minimize them. However, one should be aware that careful planning can never eliminate all risks. Despite thorough planning, there will always be a risk of losing or destroying evidence due to unexpected and emerging issues on the search scene. There might also be a risk of directly or indirectly harming people because of the investigative actions, regardless of careful planning and caution.

Prior to any action, the investigator should consider the ethical aspects of the planned actions. The goal does not justify any means, and the investigation team must carefully assess whether the actions could pose a risk of harm or injury for the victim, third parties, or the investigation team. The desire to obtain evidence of high value to the investigation should always be carefully assessed against the necessity of stopping a victim from being abused or harmed.

Technical skills of the suspect should be considered to assess the risk of evidence being hidden or destroyed. Suppose the suspect has advanced knowledge and experience. In that case, the data may be hidden in more sophisticated ways, or he or she may have planned how to effectively delete or destroy information if necessary.

The risk of violence or harm should also be assessed. Does the suspect have a history of violence? Does he or she carry a gun? Sometimes, a criminal may be armed and prepared to use violence to protect him/herself if attacked. If there is a risk of violence, the search team would need assistance from trained personnel to secure the search scene prior to performing the search.

When conducting cyber investigations, one must always consider whether the actions affect other forms of evidence. For example, a seized digital device might carry potentially relevant traces such as DNA or fingerprints, which may easily be contaminated or destroyed if they are not appropriately handled.

Before carrying out the planned actions, someone outside the investigation team should be consulted for a second opinion. This person should consider risks the investigation team might have overlooked or paid too little attention.

Example 2.4: Impact and risk assessment stage

Case example – Call4U

Impact and risk assessment stage:
The investigation has led to the following information:

The emails to the Call4U employees were routed through an anonymization service and are untraceable. However, the investigation uncovered that the Call4U business strategy information was sent to an email address outside the business domain. Basic subscriber information (BSI) was requested and revealed the IP address used when the account was created. Through search in similar case files, the NC3 revealed that the IP address had been identified in another case and was linked to a person named Mr. X. According to the information systems available to the investigation team, Mr. X lives in a house with a wife and two young children. He works as a programmer at the telecom company MoreCalls. Mr. X is not registered with any weapons and is not known to be violent.

The investigation team starts planning the arrest of Mr. X and a search of his house.

The IP address is checked, and other investigation activities are carried out to examine whether other people than Mr. X could have used the IP address. The only person was the wife, who works as a hairdresser in a local beauty salon.

The plan is now to arrest Mr. X and search his house and office for evidence. An arrest and search warrant are obtained from the court. Mr. X is a technically skilled person, and it is necessary to take precautions to avoid an opportunity to destroy evidence. The team is put together with experts on networked technology, digital forensics, and criminal investigation.

The search is planned to start when the wife and children have left for school and work to minimize the harm on innocent third parties.

A colleague was consulted about the plans. She advised the investigation team to also consider searching Mr. X's office and car since relevant evidence could be located there and would otherwise probably be lost.

2.5.5 Action and Collection

The action and collection stage aims to obtain relevant and credible information that may shed light on the case circumstances, support or refute the investigative hypotheses, or form the basis for new hypotheses.

Definition 2.8: Action and collection stage

The key tasks of the action and collection stage are:

- Obtain control over the search scene.
- Get oversight over the search scene and assess whether a change in plans is necessary.
- Prioritize the evidence collection.
- Collect the evidence in compliance with a forensically sound digital forensic process.
- Document all actions (photo, video).

The planned actions' scope might sometimes be limited, while other times broad, comprehensive, lengthy, and complex. Regardless of this, it is of great importance that the planned actions are appropriately managed and coordinated.

When entering an online or real-world search scene, the highest priority should be establishing control and oversight. Evidence integrity should be preserved to ensure that it is not altered, damaged, or lost prior to collection. When the oversight is established, the necessity of changing or adjusting the plan should be considered since new and unexpected information on the search scene may require a revised approach.

Suppose the suspect is present at a physical scene. In that case, there is a chance that he or she will try to destroy any incriminating evidence and should therefore be kept away from the items that potentially carry such information. The suspect should not be allowed to use any communication devices since they may be used to notify others or alter or delete remotely accessible evidence. If possible, an interview should be conducted with the suspect to obtain information about usernames and passwords. The initial interview should be done in compliance with the law and relevant policies and primarily aim for information about the network, relevant digital devices, Internet/cloud services, or accounts.

The scene should be searched systematically and thoroughly for digital evidence. A record with accurate information about the time of seizure, the person conducting it, and the state and context of the evidence should be created. Before moving the evidence, the context and condition in which it was found should be documented with a photo or video.

The evidence collection should be done in compliance with the original plan. However, the plan might need to be revised if something unexpected happens. The investigation team should ensure that any changes of plans are decided by competent personnel and documented in the decision log.

The collection from the search scene should be done in compliance with the principles of a sound digital forensic process (Casey, 2011; Flaglien, 2018; McKemmish, 2008). The investigator should strive to maintain evidence *integrity*. When preserving integrity is not possible, for example, when network traffic is captured, *transparency* about how the evidence was handled, changes to the evidence, and the implications of these changes should be documented.

A cryptographic checksum should be calculated when collecting the evidence, enabling the investigator to demonstrate that the evidence has remained unchanged from the point of collection to the presentation in court. A record documenting the *chain of custody* for the evidence should be created and continuously updated. When seizing evidence online, the details about the performed actions should be logged. If possible, the operation could be recorded by continuous screen capture. This provides transparency about how the information has been collected and serves as a backup if something unexpected occurs that affects the evidential value of the collected information.

Digital evidence is vulnerable, and the investigation team should take precautions to prevent damage caused by physical impact, vibrations, magnetism, electrical static, or significant variations of temperature and humidity.

In cyber investigations, the collection is not always done by the investigation team. Due to legal or sovereignty issues, where traces are stored on servers outside the investigation team's jurisdiction, external parties might do the evidence collection. The need for reliance on external parties may sometimes also be due to technological issues, where the data must be copied from a system with help from someone with specialized knowledge of the system. In such situations, the party assisting in the collection should be encouraged to follow the principles of evidence integrity, chain of custody, and transparency to obtain and preserve the best possible evidential value of the collected information.

Example 2.5: Action and collection stage

Case example – Call4U

Action and collection stage:
Mr. X was arrested on his way to work, and his house and office were searched simultaneously.
   Several objects were seized:

- A laptop from Mr. X's office
- Two laptops from Mr. X's house
- Four external hard drives and three thumb drives from a desk drawer in what seemed to be a home office in Mr. X's house
- The smartphone Mr. X carried in his pocket
- The smartwatch Mr. X was wearing
- Several notebooks and papers with possible usernames or passwords

   The information on the router was secured and seized.
   The preliminary examination of the smartphone revealed that Mr. X had access to cloud storage and several email accounts. The service providers were requested to preserve the information until the court had decided if the information should be extradited to the police.
   A court order to collect financial information from Mr. X was also obtained, and the records from his bank account were sent to the police.

2.5.6 Analysis and Integration

The analysis and integration stage aims to identify relevant information suitable to test the investigative hypotheses of the case and establish the credibility of the information in a structured and transparent manner.

Definition 2.9: Analysis and integration stage

The key tasks of the analysis and integration stage are:

- Undertake examination to make information available for analysis.
- Ensure the right competence (investigative and technical) is available for the analysis sub-stages: technical analysis, content analysis, and evidence evaluation.
- Identify traces/information that supports or contradicts/refutes the (offense level) investigative hypotheses. Detail out the hypotheses at activity, at the source levels if necessary.
- Correlate the digital evidence from different sources.
- Appoint a critical counterpart (Devil's advocate) to challenge strategies and findings.

During the analysis and integration stage, the data is processed, examined, and made available in a human-readable and understandable format (Flaglien, 2018). An examination could involve extracting data from compressed and encrypted representations, recovering deleted data, and ordering or structuring raw or unstructured data.

A house search could result in a magnitude of data, and examining all information bit by bit is sometimes impossible. The investigative hypotheses and the defined information requirement should form the scope of the analysis, guiding what is relevant to search for and what is not. The investigative hypotheses may be too general to guide the analysis, and it might be helpful to define sub-hypotheses at the activity and source levels.

During the analysis and integration stage, cooperation and coordination between personnel with investigative and technological competence are vital. The competencies are equally important to ensure effective information work during a criminal investigation. Criminal investigative competence consists of components from many different subjects and disciplines, such as law, scientific methodology, psychology, and criminology (Sunde, 2017). Technological competence could also consist of different components and specializations, such as computer science, programming and scripting, network and communications technology, computational forensics, and digital forensics.

Both competencies are crucial for testing the investigative hypotheses during the analysis stage. When planning the analysis, investigative competence is necessary to define the information requirements based on the investigative hypotheses, identify who it could be relevant to interview, and predict where any relevant physical traces might be located. Technological competence is vital to identify relevant digital artifacts and predict where they might be found.

On a high abstraction level, the analysis may be divided into two sub-stages: *Investigative analysis* and *evidence evaluation*.

The investigative analysis involves substantiating relevant traces of criminal activity within the examined data and, if any – who may be attributed to the traces. When relevant traces are identified, their credibility must also be investigated. An investigative analysis may thus be generalized to consist of:

- *Technical analysis* – primarily aimed at reconstructing activities and events based on digital traces and assessing their credibility.
- *Content analysis* – primarily aimed at making sense of and interpreting the meaning of the information in the context of the case under investigation.

Several *strategies* may be applied for the investigative analysis, for example timeline analysis, link file analysis, image content analysis, or log analysis. The result of multiple analysis strategies could be integrated to obtain a more comprehensive understanding of an event and control the credibility of the finding by cross-checking traces from different sources. Descriptions and examples of strategies are outside the scope of this chapter and are found in, for example, Årnes (2018) and Casey (2011).

Evidence evaluation is a complex analysis task that involves assessing the evidential value of traces for a particular case. Evidence evaluation involves interpretation and professional judgment and requires higher levels of technological and investigative expertise than an investigative analysis. Casey (2016) advocates that evidence evaluation requires a research foundation, formalized processes, testing, and quality oversight. The result of an evidence evaluation should be articulated as an *evaluative opinion,* which is a formalized structure for reporting expert opinion evidence (ENFSI, 2015a, 2015b). An evaluative opinion is articulated as the strength of evidence considering the defined hypotheses and the conditioning information.

A cyber investigation would sometimes require extensive and lengthy cooperation, which increases the risk of cognitive and human error due to, for example, exposure to task-irrelevant contextual information. To ensure that evidence consistent with innocence or mitigating circumstances in terms of guilt is searched for, a designated person should be given the role of the critical counterpart and challenge the investigation strategies during the investigation as well as the results. See also Section 2.6.3 for cognitive biases and relevant bias minimization measures.

Example 2.6:   Analysis and integration stage

Case example – Call4U

Analysis and integration stage:
Personnel with sufficient expertise secured and examined the digital devices in compliance with the principles from the digital forensic process. The information is now ready to be analyzed. The hypotheses and related information requirements are used to predict the location of relevant traces and identify relevant analysis strategies for uncovering traces.

Hypotheses
H1 Spear phishing attack against Call4U

- H1.1 aimed at gaining information
- H1.2 aimed at harming the company (destroying reputation)

H2 Organized spear-phishing attack – more companies were attacked
H3 An insider from Call4U has conducted or contributed to the crime
H4 No relation between suspicious emails and log activity
H5 No criminal activity has occurred

(Continued)

Example 2.6: (Continued)

- H4.1 The email attachments did not contain malicious code
- H4.2 The suspicious log activity was not a result of criminal activity

The investigation team will seek information that supports or contradicts the hypotheses. The information may also lead to new hypotheses.

The analysis of the seizure associated with Mr. X led to the following traces/ evidence:

- Activity on forums exchanging advice and links to malware designed for phishing purposes
- The malware Phishywishy was found on one of the seized thumb drives.
- There was activity on one of the laptops during the relevant time period.
- There were many Internet searches on the web domain Call4U.com and the telecom company Twocall.com.
- There seemed to be much contact with a particular email address, themrx@verysecret.com

The analysis of Mr. X's financial information and transactions led to the following result:

- Mr. X had several bitcoins transactions during the last month

Analysis of the suspicious emails and system logs at Call4U led to the following information:

- The suspicious emails had attachments that installed malware when they were opened. The malware, "Phishywishy," enabled the intruder to gain access to restricted areas in the corporate IT system, such as customer databases
- There was no sign of alteration or deletion of information during the relevant time period

Examination of themrx@verysecret.com:
The police obtained a search warrant for the email account themrx@verysecret.com. Examination of the account revealed that it was probably used as remote storage. Among the information stored on the account, the police found several spreadsheets with customer information which most likely originated from Call4U, and several documents describing the future business plan of Call4U.

Suspect interview:
The interview with Mr. X led to no further information. He did not want to provide a statement to the police and exercised his right to remain silent.

Investigation of possible phishing attack at Twocall:
The police informed Twocall that they had discovered suspicious activity toward their firm and asked them to check the system for signs of a phishing attack or intrusion. Twocall did not find any signs of anything suspicious.

The investigation team assessed the hypotheses with the findings from the investigation, with the following result:

- The information refutes H5 since credible information justifies that criminal activity has been committed.
- The information refutes H4 since credible information establishes the link between the phishing emails and the log activity.

- The information contradicts H3 since no traces indicate that an insider from Call4U has committed and contributed to the crime.
- The information regarding extensive search activity on Twocall from Mr. X's computer provides weak support of H2. However, Twocall has examined system logs and concluded that there were no traces of such an attack. Therefore, H2 is substantially weakened.
- The information supports a spear-phishing attack (H1).
  - Since there is no sign of deletion or alteration of data at Call4U or other activities to harm the company, H.1.2 is not supported by any information collected in the investigation.
  - The malware used in the phishing attack was found on a thumb drive at Mr. X's house. There were Call4U customer information and business plans on an email account he had access to, and Internet and forum activities point in the same direction. This information provides strong support for H1.11.

2.5.7 Documentation and Presentation

The documentation and presentation stage aims to disseminate information relevant to the investigation, guided by the defined scope of the remit or mandate which initiated the work. The report author/presenter should be transparent about all elements that involve uncertainty concerning the information's credibility.

Definition 2.10: Documentation and presentation stage

The key tasks of the documentation and presentation stage are:

- Disseminate in a manner that is understood by the recipient (for example the judge or juror).
- Be transparent about known limitations and errors to avoid misinterpretations.
- Make a clear distinction between findings and opinions about findings.
- Accurately present the evidence context.
- If evaluative reporting is performed, ensure compliance with the guideline for presenting evaluative opinions (ENFSI, 2015a).
- Assess the quality of the documentation through peer review.

Digital evidence is often perceived as more objective and credible than evidence based on human perception, such as witness statements (Holt *et al.*, 2018; Van Buskirk & Liu, 2006). However, since people play active and essential roles in the cyber investigation process, the potential flaws and uncertainties with the digital evidence caused by cognitive and human error should be acknowledged and acted upon.

Digital evidence does not speak for itself, and data must be discovered, interpreted, evaluated, and explained by a human before it enters the modes of information, knowledge, intelligence, or evidence. The digital evidence is presented in written or oral form that makes sense in a legal context. It is on these grounds that legal decision-makers such as judges or jurors evaluate the evidence. Therefore, the presumption of digital evidence as objective and credible is not valid, and uncertainties with the evidence should always be clearly and accurately conveyed in any written or oral dissemination of the evidence.

Reports should be written clearly, concisely, and structured and comply with the applicable standard. Written reports should include all the relevant information, which includes both positive and negative findings. Negative findings (searched for and not found) could have diagnostic value for the investigative hypotheses and should be included when relevant.

The report author should pay close attention to the importance of context and the chance of misinterpretation of evidence value if the finding is presented in isolation from the context in which it was found (ENFSI, 2015a). For example, a search for "poison" in the web browser log could be perceived as suspicious considering a hypothesis about attempted murder. However, if the complete search session from the log showed that the person searched for songs by the band The Prodigy, where one of the songs was titled Poison – the finding would have little value to the case.

In an analysis report, it is crucial to distinguish between descriptions of findings – often referred to as *technical reporting* and interpretations or evaluations of findings – referred to as *opinion evidence* (see, for example ENFSI, 2015a; ISO/IEC, 2012).

When deciding the type of reporting, one must also consider which stage the case is at and who the report is aimed at. At the investigation stage, the report serves as investigative advice by documenting leads that may guide the ongoing fact-finding. At the court stage, experts could be engaged to provide evaluative opinions concerning the evidential value (see Section 2.5.6). The same methodology may also be used during the investigative stage of the case to produce preliminary evaluative opinions (Casey, 2020). At the investigation stage, reports could include either or combinations of:

- **Technical reporting**, which is the "factual" descriptions of findings based solely on the actual evidence located (and competence of the individual) during the investigation.
- **Investigative opinions**, which involve descriptions of interpretations and explanations based on the findings and the report authors' expertise.
- **Preliminary evaluative opinions**, which are evaluations of evidential value, articulated as strength of evidence considering the defined hypotheses and the conditioning information.

At the court stage, the expert should report the result as fully evaluative opinions. For details about how an evaluative opinion should be formed and articulated, see ENFSI (2015a).

The following should, as a minimum, be considered to include in the analysis report:

Formal details

- Unique case identifier
- Details about the laboratory conducting the analysis
- The name and professional experience and competence of the analyst (relevant to the investigative task)
- Date for when the submitted material was received, name and status of commissioning party
- Date for when the report was signed

- List of submitted items with a description of condition and packaging
- Details of contextual information received with or in addition to the items
- The mandate, hypotheses, and information requirement guiding the analysis
- Whether the report is preliminary or final

Information about the investigation/analysis process

- A detailed description of tools and methodologies used in the analysis
- Whether the examination has affected the item's original state
- A description of applied strategies for mitigating technical or human error, for example dual tool verification of results or peer review of the report
- Details of known limitations in the analysis process
- A description of any items excluded from the analysis and the reasons why
- A description of items, or parts of items that are not returned to the commissioning party, and why

Terms and glossary

- If timestamps are included – time zone identification and an explanation of how the timestamps in the report should be interpreted
- If technical terms are used, a glossary with explanations of the terms

The case-relevant findings

- A factual presentation of the findings (positive and negative findings)
- Investigative opinion (interpretation or explanation of the findings)
- User or device owner identification
- The evaluation of the results in compliance accordance with the ENFSI evaluative reporting methodology (ENFSI, 2015a) see Section 2.4.3

A summary of the result and/or conclusion

- A summary of the findings
- A conclusion as an investigative or (preliminary/full) evaluative opinion, which demonstrates the reasoning of how the findings relate to the conclusion
- A description and justification of reservations, uncertainties, and limitations concerning the findings and conclusion

2.5.7.1 Evaluative Reporting

If a preliminary or fully evaluative opinion is presented, it should follow the principles and methodology for evaluative opinions (ENFSI, 2015a). This involves presenting the strength/probability of the findings in the light of the hypotheses, as opposed to the strength of the hypothesis (ENFSI, 2015a). There has been a long-lasting discussion within forensic science about the best way to convey the probative value/strength of the evidence. Some have pointed to statistical terms and numeric presentation of likelihood ratio, and others have pointed to qualitative descriptions. A numeric scale is better for communicating a change in the probative value of the evidence in one or the other direction when new evidence is considered. The verbal scale might make more sense if the assessor is not well-versed in statistics. In the following, we stick to the qualitative descriptions. However, for those who want to explore the quantitative approach, the guidelines from the Royal Statistical Society (Aitken *et al.*, 2010) are recommended.

Example 1: A balanced presentation of findings in relation to two hypotheses are:

> *The findings from the examination support that the suspect downloaded the child sexual abuse images from the Internet (hypothesis 1) rather than that the images were downloaded by a malware program (hypothesis 2).*

Example 2: An example of a balanced presentation of findings in relation to three hypotheses are:

> *The evidence does not support the hypotheses that the suspect hacked into the computer system.*
> *The evidence provides weak support for the hypothesis that the suspect has planted evidence on the computer to make it look like another party controlled the system.*
> *The evidence provides strong support for the hypothesis that the suspect's computer was remotely controlled and that someone else hacked into the computer system by using the suspect's computer.*

Example 3: An example of the strength of hypothesis approach that should be avoided is:

> *The findings support that the suspect is guilty of downloading child sexual abuse material from the Internet.*

A presentation such as Example 3 should be avoided since the question of guilt is the court's responsibility and not the investigation team. The evaluation should instead present the evidence relevant to the question of guilt in a clear and balanced manner so that the court has the best foundation to make an assessment.

Research on the certainty expressions for forensic science opinions has revealed that a wide variety of expressions are in use, all of which have their strengths and limitations (Jackson, 2009; Sunde, 2021a). Expressions such as "cannot exclude" or "consistent with" are commonly used but can be misleading. "The examination of the material *cannot exclude* the suspect from being the source of the biological material" is a statement that contradicts the hypothesis but may easily be misconceived as a statement supporting the hypothesis. "The Internet searches on the terms knife and decomposition are consistent with the suspect planning to kill his wife" might be a reasonable opinion. However, there may be other explanations of the searches that are not linked to the homicide, which should be investigated and documented.

2.5.7.2 Peer Review

Peer review is a critical measure to safeguard the quality of the investigative results. Page *et al.* (2018) describe five different peer review concepts, further developed into the seven-leveled Peer Review Hierarchy for Digital Forensics (Horsman & Sunde, 2020). The levels consume the levels below, which means that, for example, a Conceptual review also includes a Sense review, Proof check, and Administrative check (Figure 2.4).

The first three levels are general and centered on other aspects than the investigative or evidential information in the report. The *administrative check* focuses on whether the practitioner has performed the agreed investigation of all the exhibits due to the commissioning party's requirements. The *proof check* centers on grammatical and

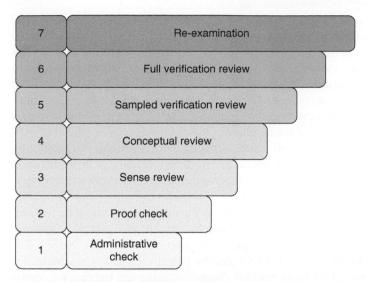

Figure 2.4 The peer review hierarchy for digital forensics (adapted from Horsman and Sunde (2020)).

spelling errors and whether terms and abbreviations are used consistently. The *sense review* examines whether the report is clear and understandable and whether any technical terms are sufficiently explained.

The following levels focus on the information concerning the investigative procedures and the Reported results. A *conceptual review* examines the report, with a particular focus on its logical and scientific underpinning. The review is based on five fundamental principles (Sunde & Horsman, 2021), which are summarized as:

- Balance: Whether the relevant hypotheses are defined, and the result is presented considering these.
- Logic: Whether the inferences, assumptions, and interpretations are valid according to how they are explained and justified.
- Robustness: Whether the scope of the examination has provided sufficient basis for the conclusion.
- Transparency:
  - Whether tools, methods, and procedures are adequately documented.
  - Whether the findings are accurately presented in terms of what they are and the context they were found.
  - Whether any reservations, uncertainty, and limitations concerning the tools, methods, procedures, findings, or conclusions are conveyed.
- Compliance with ethical/legal regulations:
  - Whether the examination and reporting were done in compliance with the applicable ethical guidelines and law.
  - Whether the presumption of innocence was safeguarded through a search for exculpatory evidence or traces indicating mitigating circumstances for the suspect.

However, the analysis report provides an insufficient basis for verifying the results. The scope of a *Sampled Verification Review* goes beyond the analysis report and involves accessing the evidence file to verify a representative sample of the findings. The reviewer should use a different tool than the initial examiner to minimize tool interpretation error. A *Full Verification Review* entails verification of all the reported findings. A limitation of level four to six is that, since knowing who performed the initial analysis, the findings, and the conclusion might bias the reviewer due to anchoring and expectancy effects. These sources of bias are mitigated in the top level of the Peer Review Hierarchy. A *Re-examination* entails that an independent investigator performs examination, analysis, and documentation on the evidence file a second time without knowing who performed the initial investigation and the results.

In order to uncover errors and uncertainties, analysis reports should, at a minimum, undergo a Conceptual review (which includes the levels below). The Verification Reviews and Re-examination are resource-intensive but should be implemented regularly to uncover both case-specific and systemic errors, identify the need for quality improvement, and enable organizational learning. A Re-examination could be relevant in situations such as when the seized data is outside the practitioner's standard competency framework or when it is a rare case type to the practitioner (ENFSI, 2015b). The decision to perform a Re-examination could also be based on a risk assessment, for example when the uncovered evidence has high diagnostic value and the case's outcome has severe consequences for the suspect. An example of such a situation is when the suspect is charged with murdering her husband, and the digital evidence is essential for proving that she is guilty of the crime.

An expert opinion may be of great value in court but could easily be misconceived as factual information. Hence, in the same manner, as in written reports, it is vital to be clear about whether one is presenting findings or opinions about the findings when presenting orally in court. Research on wrongful convictions has shown that misleading forensic evidence does not necessarily relate to the evidence itself but instead the presentation of its relevance, probative value, and validity (Smit *et al.*, 2018). A common misleading aspect of forensic evidence was to present the belief in a hypothesis versus another hypothesis disproportionately to the actual probative value of the evidence, for example, by overstating to what degree the evidence supports one hypothesis versus another.

Judges and jurors tend to consider forensic evidence as quite reliable (Garrett & Mitchell, 2016). To avoid misinterpretations of evidential credibility, it is important to be transparent about the known limitations and uncertainties concerning the digital evidence. For example, if the expert had to deviate from an established procedure, the deviation should be explained and justified, and the measures to ensure quality should be elaborated. This will allow the court to make an informed judgment about the probative value of the evidence.

Example 2.7: Documentation and presentation stage

Case example – Call4U

Documentation and presentation stage:
Several reports were written during the investigation:

- Reports about the search and seizure at Mr. X's house and workplace
- A report about the attempted interview of the suspect

- Report about the analysis of the seized material. This report contained a combination of technical reporting and evidence evaluation. The narrative of the conclusion was written in a balanced way, describing the probative value and the evidence considering the relevant hypotheses:

Conclusion:
The digital forensic findings

- Provide strong support for the hypothesis that a spear-phishing attack was performed against Call4U aimed to gain customer information (H1.1).
- Do not support that the phishing attack against Call4U was aimed to harm the company (H1.2).
- Do not support that this was an organized spear-phishing attack against several companies (H2).
- Do not support that an insider from Call4U conducted or contributed to the crime (H3)
- Do not support that there is no relation between the suspicious mail and log activity (H4)
- Do not support that no criminal activity has occurred (H5)

The analysis of the seized evidence

- A peer review was performed on the analysis report.

2.5.8 Evaluation

The evaluation stage aims to learn from both mistakes and successes, develop measures to minimize errors and maintain good practices.

Definition 2.11: Evaluation stage

Key tasks of the evaluation stage are:

- Assess: What went well, and why?
- Assess: What did not go as planned, and why?
- Define learning points, develop measures, and a plan for implementation.

Learning from successes and mistakes in every investigation is essential to sustain or improve the quality of both process and results. All those involved in the investigation will possess valuable experiences and should contribute to the evaluation.

The investigation plan and the decision log are two key information sources that should be used actively during the evaluation. Another vital source of information is the verdict or other court documents, where assessments of the relevance and credibility of evidence (and presentation of evidence) are described. When errors are uncovered, attention should be directed toward these documents to identify what caused the errors instead of looking for someone to blame. Identification of plausible causes is of great value to future learning, and they should be transformed into learning points and concrete measures.

However, evaluation should not only be directed toward avoiding future mistakes but also be aimed at ensuring that fruitful actions and successful measures are being repeated in the future. A plan for implementing measures encompassing all relevant organizational levels should be developed to avoid repeating errors and ensure that good practices are continued in future investigations.

Example 2.8: Evaluation stage

Case example – Call4U

Evaluation stage:
After the investigation of the incident at Call4U was completed, the investigation team had a meeting where they discussed each stage of the investigation, and aimed to extract learning points:

Assess: What went well, and why?

– They provided a broad set of hypotheses, which enabled the investigation team to look for incriminating evidence and evidence of innocence.
– The team had the necessary competence to investigate the incident in a high-quality and effective manner.

Assess: What did not go as planned, and why?

– The team had assumed that Mr. X traveled by car to work and had planned to search the car immediately after the arrest. However, Mr. X traveled by bus to work, and his wife used the car since she drove the kids to school every morning. Therefore, the team had to postpone searching the car and acknowledged that they should have investigated this issue better during the search planning.
– The team assumed that Mr. X was willing to give his statement to the police and therefore brought him back to the police station. At the police station, Mr. X stated that he had decided not to talk to the police. The team agreed that they should have attempted a preliminary interview immediately after the arrest, after informing the suspect of his rights.

Define learning points, develop measures, and a plan for implementation.

– Better research in advance of a house search.
– Bring an audio recorder to the search scene and aim at a preliminary interview of the suspect immediately after the arrest.

2.6 Cognitive and Human Error in Cyber Investigations

Common to criminal investigation and science is that they are built upon trust and honesty and that the human factor plays an essential role in defining and achieving the objective. Humans are impressionable, impulsive, and subjective, which means that every person involved in an investigation, regardless of competence or experience, is prone to unintentional human error. Within the domain of digital forensics, there has been a movement from perceiving tools and technology as the main instruments in the digital forensic process toward a greater acknowledgment of *the human* is just as a vital instrument for examining digital evidence (ENFSI, 2015b; Pollitt *et al.*, 2018; Sunde & Dror, 2019).

This section describes some of the limitations and sources that may cause cognitive and human errors and some relevant countermeasures.

2.6.1   Cognitive Factors

The human brain is not a computer, and it has limited capacity when perceiving, processing, and recalling information. The brain uses several strategies to cope with this, such as chunking, selective attention, and top-down processing (Sunde & Dror, 2019). *Chunking* means connecting individual pieces of information, so they appear as one unit within our mental representation. *Selective attention* refers to our attention to specific pieces of information while ignoring others. *Top-down processing* refers to a set of cognitive processes driven by the information already in our brains. In order to make sense of incoming information, it is assessed in light of context such as our knowledge, expectations, hope, fears, and prior experiences (Lindsay & Norman, 1977; Zapf & Dror, 2017). These strategies happen unconsciously and help us to handle information efficiently in everyday life. Their downside is that they degrade our impartiality and cause biases.

These mechanisms, along with several others, play important roles when developing expertise. They allow the expert to perform their tasks quickly and effectively. Paradoxically, the same mechanisms entail computational trade-offs that sometimes degrade the performance since they restrict flexibility and control. The expert may lose attention to detail and suffer from tunnel vision (Dror, 2011). Hence, and contrary to what many intuitively would believe, in various ways, an expert may be more susceptible to bias than novices due to how the brain develops when becoming an expert (Dror, 2011).

2.6.2   Cognitive Biases

Cognitive biases can be described as systematic errors in thinking and decision-making, which lead to systematic deviations from what may be assumed (on logical grounds) as rational behaviors (Granhag & Ask, 2021). Cognitive biases are the effect of internal and external sources impacting cognition and decision-making. Several cognitive mechanisms, referred to as *heuristics*, have been identified as contributing factors for biases. Heuristics may be described as mental shortcuts or rules of thumb characterized by fast and intuitive thinking (Tversky & Kahneman, 1974). The list of heuristics and biases described in the literature is quite extensive. However, the focus here will be on biases that we consider particularly relevant to those involved in cyber investigations.

The anchoring effect means an initial value is established and serves as an anchor from which all the information received later is adjusted. Anchoring entails that the order of the information we receive affects the meaning and value of the information. The information we receive first is given more weight than what we receive later (Tversky & Kahneman, 1974). Charman and colleagues found that DNA evidence had a more significant impact on the jurors' evaluation of the evidential weight when presented after (as opposed to before) the suspect's alibi (Charman *et al.*, 2016). In the context of an investigation, anchoring may limit the ability to change perspective, and one can remain at a perception of what kind of matter one is facing, even if information later indicates something else.

The availability heuristic makes people overestimate the likelihood of an outcome based on our limited ability to recall similar instances of the same outcome (Tversky &

Kahneman, 1974). For example, when a new case comes in, and the investigator recalls several similar cases that led to convictions, the availability heuristic may affect the ability to safeguard the presumption of innocence principle. A tendency to believe that a suspect is guilty is themed *guilt bias* (Ask & Granhag, 2005).

The representativeness heuristic leads to erroneous likelihood predictions (Tversky & Kahneman, 1974). We are insensitive to base rates and sample sizes and tend to overestimate coincidence as regularity (Helstrup & Kaufmann, 2000).

In the process of making sense of what is observed, a hypothesis will often evolve subconsciously. The *confirmation bias* (Nickerson, 1998) is the tendency to:

a) seek information that supports the hypothesis,
b) interpret new information in relation to the hypothesis,
c) interpret ambiguous or neutral information as supportive of the hypothesis,
d) ignore or explain away information that contradicts the hypothesis, and
e) assign little weight to information that does not support the hypothesis.

Suppose a person is suspected of committing a criminal act. In such a situation, the confirmation bias may lead to an aggregated focus on finding information consistent with the guilt hypothesis, with a tendency to overlook information indicating innocence or mitigating circumstances.

Groupthink is a bias one should be particularly aware of in cyber investigations, where close cooperation is often necessary. This bias implies an attraction toward consensus and unity and a reluctance to think critically and challenge the dominant opinion or theory of the group (Janis, 1972).

Several contextual factors influence the observations and conclusions of those involved in an investigation. The sources that can bias observations and conclusions in forensic work have been categorized in an eight-leveled taxonomy (Dror, 2020). The taxonomy moves from general sources related to human nature, environment, culture, and experience to case-specific sources. The case-specific sources of bias are related to irrelevant case information, reference material, case evidence, and the data. The contextual information as a biasing source introduces a dilemma in the cyber investigation process: On the one hand, in order to ensure the correct competence is available at the right time, close cooperation is necessary. On the other hand, close cooperation increases the risk of groupthink and bias caused by sharing task-irrelevant case information.

When biases emerge, there is a risk that they propagate into the further investigation process. A *bias cascade effect* may occur, which means that the bias cascades to other stages of an investigation (Dror *et al.*, 2017). For example, when traces at the crime scene are perceived to indicate guilt for the suspect, this belief may, in turn, influence the observations, interpretations, and decision-making during the analysis of digital evidence (Sunde & Dror, 2021). There is also a risk of the *bias snowball effect*, which means that the bias grows stronger because several sources of task-irrelevant information being integrated and influencing each other (Dror *et al.*, 2017).

For example, a witness identifies the suspect in a line-up (irrelevant information 1), which makes the detective believe that the suspect is guilty of the crime. The detective informs the digital forensic examiner about his belief concerning the suspect's guilt (irrelevant information 2), but that more evidence is necessary to prove it (irrelevant information 3). This may influence the digital forensic examiner to – unknowingly and unintentionally – look for traces of guilt among the digital evidence and overlook or explain away traces pointing toward innocence.

The knowledge of the "need for more evidence" may increase the motivation or perceived pressure on the digital forensic examiner to find information consistent with the suspect's guilt. The bias snowball may also influence the evaluation of evidence, where ambiguous evidence is interpreted as consistent with a particular hypothesis. Eventually, the bias snowball may lead the digital forensic examiner to assign an inflated probative value of the evidence when presented in the analysis report or court.

To ensure high-quality and effective cyber investigations, countermeasures aimed at minimizing cognitive and human error should be implemented. Relevant countermeasures are discussed in the next section.

2.6.3 Countermeasures

First and foremost, to understand how observations and conclusions may be biased, basic psychology theory should be included in the training of all involved in cyber investigations. However, knowing about the risk of bias is not a sufficient measure to minimize it to several common incorrect beliefs about bias.

Common fallacies described by Dror (2020) are, for example, to believe that bias is an *ethical issue* and only happens to people with bad morals, or that bias is a *competency issue* that only happens to novices who do not know how to do their job correctly. The *bias blind spot* entails thinking that one is "the exception" who is immune to bias due to extensive professional expertise and that other experts are biased. Another fallacy is the *illusion of control*, believing that one may counter bias by mere willpower. However, particularly relevant to cyber investigations is the fallacy of *technological protection* from bias, where one believes that advanced technology, instrumentation, automation, or artificial intelligence protects from cognitive biases. These can reduce bias, but the human is "in the loop" by designing, building, and using the system as well as interpreting the information produced by the system. However, it is critical to acknowledge that technology can also be a biasing source, for example, by how it orders the information presented to the cyber investigator, what information is highlighted or downplayed, or how the system presents the information's context.

Training is a crucial measure for changing incorrect beliefs about bias. The training should be practical and related to cyber investigation casework, allowing the participants to experience how bias could emerge and impact judgments and decisions in a typical work situation.

To conduct the examination in a cyclic process of forming, testing, and eliminating *multiple*, and preferably *competing,* hypotheses is a relevant strategy to mitigate confirmation bias since the investigation team then is forced to consider more than one hypothesis. To safeguard the objectivity and minimize a possible guilt bias, as a minimum, the set of hypotheses should include a hypothesis implying innocence for the suspect. This measure is integrated into the ICIP and would, if conducted properly, contribute to minimizing bias.

Although every stage of the ICIP involves a critical assessment of some sort, groupthink is a strong bias. The measure of appointing a critical counterpart (Devil's advocate) dedicated to challenging strategies and results is considered as a relevant bias minimizing measure and is therefore implemented in several of the ICIP stages.

Though measures such as compartmentalizing the work (Dror, 2014) or exposure control (Dror *et al.*, 2015) are relevant to reduce contextual bias, these are not always practical or possible in cyber investigations. Thus, a measure that should be addressed in every investigation is some level of *context management* (Dror, 2014). The lead investigator should have complete oversight of the case information and pay close attention to what information is shared with whom. However, as already mentioned, the need for close cooperation and information exchange will complicate this task. If the teamwork involves extensive information sharing, an important measure is at least to minimize the sharing of clearly task-irrelevant biasing information, such as one's beliefs, judgments, feelings, sympathies, and antipathies. Examples of irrelevant biasing information are sharing feelings about the crime or expressing sympathy for a victim to others in the investigation team. It could also be to share beliefs about who committed the crime or how one perceives the trustworthiness of a witness.

One of the most important lessons to take from this section is to remember that no one is immune to bias and that bias and human error can never be eliminated. Therefore, it is essential to acknowledge own and others' cognitive limitations and welcome measures for bias mitigation to safeguard a high-quality cyber investigation.

Example 2.9: Guccifer hacking of social media accounts

The Romanian hacker under the alias Guccifer was arrested in 2014 for hacking several AOL, Yahoo!, Flickr, and Facebook accounts in 2012–2014. The accounts belonged to several Romanian officials and approximately 100 Americans, among them current and former U.S. politicians and government officials. The victims included an immediate family member of two former U.S. presidents, a former member of the U.S. Cabinet, a former member of the U.S. Joint Chiefs of Staff, and a former presidential advisor.[1] Guccifer publicly released his victims' private email correspondence, medical and financial information, and personal photographs.

One of the victims of Guccifer was Dorothy Bush Koch, the sister of former president George W. Bush. After hacking her AOL account, Guccifer leaked images of George W. Bush in hospital and a self-portrait painting of Bush naked in the shower. He also hacked the email account of Sidney Blumenthal, a former aide to President Bill Clinton, and distributed private memos from Blumenthal to Secretary of State Hillary Clinton involving the Benghazi attack (2012) in Libya.

The investigation showed that the hacker had no particular computer expertise or fancy equipment. He used the simple technique of finding private information about his victims online and then using this information to guess the correct answers to security questions. At the time of his arrest in 2014, Guccifer was an unemployed taxi driver.

Guccifer was sentenced to four years in prison in Romania for the crimes conducted against Romanian officials. In March 2016, Romania approved an 18-month temporary extradition to the United States, and he was surrendered to U.S. authorities. Guccifer was sentenced to 52 months in prison for the crimes involving American victims.

---

1 https://www.justice.gov/usao-edva/pr/romanian-hacker-guccifer-pleads-guilty-computer-hacking-crimes

2.7 Summary

A cyber investigation is a form of construction work where digital traces are transformed into evidence through a series of investigative activities. In this chapter, the ICIP was presented as an integrated process for cyber investigations, which integrates the criminal investigation process involving testimonial and tangible evidence with the digital forensic process where digital evidence is handled. The two processes are integrated from the start of the investigation and require close cooperation between personnel with investigative and technological competence at every stage. The Cyber Investigation Queries serves as a useful thinking aid in the ICIP and is used to make sense of the case information, develop hypotheses about the specific event under investigation, and predict where relevant traces may be located.

Due to the extensive cooperation required in cyber investigations, it is vital to pay close attention to the risk of human error. This chapter has presented several measures to prevent cognitive and human error during cyber investigations, such as a scientific approach to the investigation by forming and testing hypotheses, by actively seeking second opinions from a critical counterpart outside the investigation team, by ensuring that the investigative tasks are only conducted by personnel with sufficient competence, and quality control measures such as peer review are conducted. Evaluation plays a significant role as a quality measure in the ICIP and is integrated as a sub-process in every stage, as well as a top-down evaluation process in the final stage of the investigation.

2.8 Exercises

1 The ICIP is an integration of two processes. Which are they?

2 The Cyber Investigation Queries play an essential role in the ICIP. What is the purpose of using this model in cyber investigations?

3 What is the ABC rule?

4 In the Investigation initiation stage, what should be the first task after identifying the first hypothesis, and why?

5 What are the important elements of an investigation plan and a decision log?

6 What are the key tasks of the Impact and risk assessment stage?

7 What is the aim of the Action and collection stage?

8 In the Documentation and presentation stage, what is the difference between technical reporting, investigative opinions, and evaluative opinions?

9 In the Documentation and presentation stage, what does it entail to present the findings in a balanced manner?

10   What are the objectives of the Evaluation stage?

11   Awareness is not a sufficient measure to eliminate bias. Why?

12   What is the confirmation bias, and how does it manifest itself in an investigation?

13   What is the main cause of "bias cascade" and "bias snowball," and what is the difference between these two bias dynamics?

14   Describe the relevant bias mitigation strategies in cyber investigations.

3

Cyber Investigation Law

*Inger Marie Sunde**

The Norwegian Police University College, Oslo, Norway

This chapter presents legal norms relevant to cyber investigations performed for law enforcement, preventive, and strategic purposes. The mandates of the police organization and the national intelligence service (NIS) differ, and within the boundaries of their respective mandates, the purpose of cyber investigations may differ. All of this impacts the legal framework applicable to the concrete situation. Furthermore, there is a crucial distinction between the legal powers of public agencies and the private sector investigator. The chapter also presents jurisdiction issues and legal mechanisms for international cooperation in criminal cyber investigations and evidence exchange. The presentation builds on *Cybercrime Law* (Sunde, 2018) in *Digital Forensics* (Årnes, 2018).

3.1   Cyber Investigation in Context

In Section 1.3 (Definition 1.5), the term cyber investigation is defined as "a systematic collection, examination, and evaluation of all information relevant to establish the facts about an Internet-related incident or an alleged cybercrime and identify who may have committed or contributed to it." The definition defines the concept (and activity) this book is about. The present chapter elaborates on the different contexts in which such processes may be conducted. It will become clear that the *mandate* of the cyber investigator, combined with the *purpose* of the investigation, determines the legal rules applicable. For the process to be lawful when applied in practice, it must be adapted to the legal framework relevant to the concrete situation.

First, we describe cyber investigation relative to the *mission* of the organization for which the cyber investigator works (Section 3.2). Here a crucial distinction goes between public organizations endowed with special legal powers and the private sector cyber investigator (PCI), who is in the same legal position as any other citizen. The public organizations we address are the law enforcement agency (LEA), the police, and the

---

* Professor Inger Marie Sunde, PhD and LLM.

internal and external national security and intelligence services (collectively referred to as NIS in this book).

By LEA, we mean the function in the police mandated with criminal investigations for the purpose of holding criminals to justice. Cyber investigators working for the LEA will often interact with the public prosecutor's office and/or an investigative judge. Together they are responsible for conducting criminal investigations, bringing the criminal charge, and prosecuting, thus enforcing the law.

The very broad notion of "cyber investigation" applied in this book implies that its aim may be manifold; it may be to *investigate* in order to prosecute a crime that was committed, to *directly prevent* cyber-crime and cyber-enabled crime from being committed, and to gain information for *intelligence* purposes. The LEA is not tasked with *direct* crime prevention, that is, concrete preventative initiatives aimed at preventing a crime from happening. Criminal prosecution and punishment are deemed to have preventive *effects,* meaning that they have an *indirect* adverse impact on possible new crimes (see (Muir, 2021), about direct and indirect prevention of crime and harm). The police organization has many mandates apart from criminal investigation, including tasks relating to *direct crime prevention*. In relation to cyber investigations, *intelligence activity* is relevant as well. When these are the sole aims of the cyber investigation, other functions of the police organization than the LEA are in charge.

The role of the police and the LEA must further be distinguished from that of the NIS. The NIS often works closely with the police and the LEA, with NIS often performing the covert part of a cyber investigation and the LEA taking over the further prosecution of the suspects. The different mandates of the LEA, NIS, and the police are described in Section 3.3.

We then turn to issues concerning jurisdiction and international cooperation (Section 3.4) and cyber investigation methods in the context of the fundamental rights to a fair trial and to privacy (Section 3.5). After that, we revert to the PCI and address data protection (General Data Protection Regulation (GDPR)) and privacy protection in the context of labor rights (Section 3.6). Routinely, the chapter ends with observations regarding the way ahead (Section 3.7), a summary (Section 3.8), and exercises (Section 3.9).

3.2   The Missions and Some Implications to Privacy Rights

The joint mission of the police, the LEA, and the NIS is to protect citizens, society, and national security against crime and harmful incidents. To be able to fulfill their mission, the organizations may be empowered to make use of so-called *coercive measures* that interfere with fundamental rights to personal liberty, property, or private life (see section 3.2.3.6 in Sunde (2018)).

3.2.1   The Police, Law Enforcement Agencies, and National Security Service

In relation to cyber investigations, the use of *privacy-invasive* methods is particularly relevant. Such methods typically interfere with the right to respect for one's family life, social relations, home, geographical movements, private communications, and data protection rights.

Traditionally the legal competence to make use of privacy-invasive methods is limited to criminal investigations. Criminal investigations primarily fall within the remit of the mandate of the LEA. The competence is not afforded to the police in its direct preventative function because every citizen has a right *not* to be subjected to any systematic attention from the police when s/he respects the law.

3.2.2 Reasonable Ground to Open a Criminal (Cyber) Investigation

To become a person of interest in a criminal investigation may be uncomfortable and burdensome. Recognizing this, the law first creates a threshold for *opening* a criminal investigation, usually demanding *"reasonable ground"* to believe that a crime has been committed (this includes criminal attempt and ongoing crime). The condition implicates that criminal (cyber) investigation is a *suspicion-based* activity. It is not necessary that the suspicion concerns a known suspect, but, as a minimum, the criminal act must be somewhat concretized by facts. Reasonable ground may be established by a formal report of a crime typically filed by the victim. It could, for instance, be a corporation that has suffered a cyber-attack. Reasonable ground may also be established by concrete factual circumstances which otherwise have come to the agency's knowledge, for instance, through patrolling and systematic intelligence work. Criminal investigation of organized crime is often based on initiatives taken by the police organization itself. Such crime is known as "victimless crime" because no one dares or takes a personal interest in informing the authorities about the unlawful activity. Hence the police organization, including the LEA, must take unilateral action. However, by the threshold "reasonable ground," the citizens are legally protected from arbitrary investigations initiated by a gut feeling or zealousness of a cyber investigator.

3.2.3 The Legal Framework(s)

The previous section described the threshold for opening a regular criminal (cyber) investigation, for which the LEA is in charge. The legal framework for this is mainly *criminal procedural law.* The powers and conditions for the use of coercive measures are also laid down in this framework.

As explained, cyber investigations may be performed outside the context of a criminal investigation as well. This falls outside the remit of the LEA, but other functions of the police organization or the NIS could engage in this. The police organization may, for instance, have an online patrolling function openly present in social forums or be gathering information from open sources (open-source intelligence (OSINT)). Such activity is regulated by *general rules of policing* combined with the *data protection rules* of the police.

General rules of policing have typically evolved nationally, whereas the data protection rules, at least in Europe, are based on the European Police Directive (2016/680). This is the "sister" to the European GDPR (2016/679). The directive requires the police to limit the gathering of information to what is relevant and necessary for the purpose and imposes strict requirements regarding data security, traceability, data deletion, and oversight.

The Police Directive also governs criminal investigation. However, the demanding challenges of combatting serious crime and meeting the burden of proof have justified broad exemptions from the directive in this respect. Instead, criminal (cyber) investigations are afforded wide discretion to determine which information is relevant and necessary to collect.

One may note a tension between the long-lasting principles of criminal procedural law and the more recent but ever-more important rules of data protection law. This has caused some challenges experienced by European countries in the transposition of the Police Directive into national legislation. Custers (2021) further elaborates problems of this nature in relation to Dutch criminal law on evidence. To the legal expert, the challenge (and the art) is how to reconcile legal principles stemming from different traditions. Problems of this kind might be exacerbated in relation to the emerging European legal framework for artificial intelligence (AI) (see Section 3.7).

Finally, the "NIS" may be partly internally oriented (iNIS) and partly externally oriented (eNIS). Both functions may be governed by special legislation (see Section 3.3.2). In addition, data protection rules based on the Police Directive apply to the iNIS, but not to the eNIS, as external security activities fall beyond the scope of the directive.

3.2.4 General Conditions for Privacy-Invasive Cyber Investigation Methods

Methods relevant to a cyber investigation could be, for example, computer surveillance, interception of electronic communication, computer search and seizure, infiltration, systematic observation of specific individuals and infiltration (for instance, on social media), and the systematic storage and processing of personal data.

Such methods interfere with privacy rights and must fulfill certain conditions to be lawful. Personal data are doubly protected, as data protection rules apply in addition to the rules already embedded in the fundamental right to privacy (cf. Satakunnan Markkinapörssi Oy and Satamedia Oy v Finland, Judgment (GC) (2017), chapter 11.4 VIII and pp. 505–506 in Harris (2018), and chapter 16.5 in Kjølbro (2017)).

There is no international uniform way to distribute powers between the police, the LEA, and the NIS. In this section, we are not concerned about whether each of them has the competence to make use of privacy-invasive methods when carrying out their mandate. This is for the national jurisdiction to decide. The focus here concerns human rights, specifically the right to privacy. This right (and freedom) is universally applicable and must be respected and secured by every nation-state. Hence, in the following, we describe *the general conditions* for applying *privacy-invasive* methods as mentioned above. These conditions must be understood to understand the legal framework of cyber investigations.

The fundamental right to privacy is embedded in numerous international conventions and national constitutions, see section 3.2.3 in Sunde (2018). Here we refer to article 8 of the European Convention of Human Rights of November 4, 1950 (ECHR), which in the first paragraph states that everyone has *"the right to respect for his private and family life, his home and his correspondence"* (cf. section 3.2.3.6 in Sunde (2018)). At its core, this is a right to be left alone and not be subjected to any systematic interest from public authorities. As per the case law of the European Court of Human Rights (ECtHR), privacy rights stemming from the word "home" in article 8 apply *mutatis mutandi* to private corporations, see Societeé Colas Est and Others v France (2002), chapter 12.2 VIII in Harris (2018), and chapter 16.1.3 in Kjølbro (2017). The legal protection of the

ECHR article 8, therefore, extends also to the business premises of private corporations and other organizations.

According to the ECHR article 8 second paragraph, *three conditions* must be fulfilled for such interference to be lawful: the method must have a legitimate aim, a legal basis, and be proportionate in relation to the goal it seeks to attain.

The first condition – *legitimate aim* – does not pose any problem in relation to cyber investigations conducted by the LEA, police, or NIS. Pursuant to article 8, second paragraph, legitimate aims are, i.a., *"the interests of national security, public safety [and] the prevention of disorder or crime."* Although the wording may indicate that solely preventive purposes are legitimate, long-standing case law has clarified that investigation of past events – such as crimes fully completed, with a view to bringing the perpetrator to justice – is a legitimate aim as well, as we see in Camenzind v Switzerland (1997), Z v Finland (1997), chapter 12.4 VII in Harris (2018), and chapter 15.2 in Kjølbro (2017). The mission of the said organizations (protection of citizens, society, and national security) is clearly covered by the cited part of article 8; hence is legitimate.

The second condition is that the interfering (privacy-invasive) method must have a *legal basis*. As explained by Sunde (2018), page 62, this means that *"the method is adequately described and authorized in national legislation."* The legality condition has two important functions. Firstly, it secures democratic control with public authorities' use of privacy-invasive methods and interference with data protection rights. The legislative procedure involves the parliament (i.e. the legislator), which is democratically elected. Thus, the nature and scope of the powers of the police, the LEA, and the NIS have some acceptance and legitimacy in the general population. Secondly, the legislation applies equally to everyone and sets clear boundaries not to be exceeded by the agencies. This way, legislation is a guarantee against arbitrary privacy interferences by public authorities.

The third condition, *proportionality*, implies that a privacy-invasive method may be used only if there is *"a pressing social need"* ((Olsson v Sweden (No 1) (plenary), 1998), chapter 11.3.III in Harris, (2018), and in chapter 15.3 in Kjølbro, (2017)) and other less intrusive measures are unsuitable for the purpose. The law requires that the agencies assess proportionality each time the question of making use of an invasive method arises. In addition, the method must be prereviewed by a judge (excepting exigent circumstances). This applies both when the method is used covertly and when performed with the suspect's knowledge. Digital search and seizure, for instance, are methods that (always depending on the circumstances) can be useful also when performed with the suspect's knowledge. The methods nevertheless interfere with the suspect's fundamental right to privacy, so a judicial review is required before the method is activated.

Cyber investigations will often be performed covertly (see Section 3.5.2), hence perceived as a menace to democracy, cf. the landmark case (Klass and Others v Germany, 1978). To maintain control, the national control system must provide for subsequent independent oversight too. Covert methods are therefore regularly subject both to prior judicial review and subsequent oversight.

Control procedures are a prerequisite for accountability yet can be effective only if adequate and reliable documentation exists. The documentation should be clear and concise and made as close in time to the cyber investigation as possible. It should disclose who made decisions, the basis for decisions, how a method was carried out, by whom, and how information was collected, analyzed, stored, and eventually deleted, made inaccessible, or otherwise protected from unlawful exposure. Any alterations to (digital) evidence should be disclosed, the cause and any effects on the reliability of the evidence should be explained (cf. pages 67–68 in Sunde (2018)). The cyber investigator is responsible for producing sufficient and reliable documentation of the activities carried out during the investigation. The leaders of the organization should make technical equipment available with functions supporting the documentation needs.

In sum, the law forces the government to have in place crime prevention and law enforcement systems that respect the rule of law and include safeguards and procedural guarantees necessary for placing trust in the system.

3.2.5 The Private Sector Cyber Investigator

A private sector investigator (PCI) is a professional who works in a private capacity, either as a hired consultant, e.g. for corporations experiencing cyber security incidents, or employed as an in-house security expert. The PCI does not have powers corresponding to those described above. The PCI is a private citizen who, like other citizens, is free to do what is not prohibited. In addition, the PCI must comply with the terms agreed with the principal and the general legal framework relevant to the profession. Data protection law, such as the GDPR, is probably the most important set of rules in this respect, but issues concerning employees' privacy rights may be practical as well.

Many of the methods described in the previous section are, *as per their nature*, criminal offenses, i.e. unauthorized access, interception of electronic communication, data and system interference (cyber vandalism), and production and possession of unlawful computer devices and malware. The background of the criminalization is explained in Sunde (2018, s. section 3.3.3). The PCI is regarded as any other citizen in relation to criminal law. It follows by implication that the methods are not legally available to the PCI.

This is different from cyber investigators working for the public agencies we have addressed. In that case, the law may give priority to crime prevention and law enforcement at the cost of citizens' privacy rights. Hence public agencies' cyber investigators are not criminally liable when making use of the methods, that is, on the condition that they operate within the limits of the special legal powers given to them by the legislator. As explained, the legal power is contingent on clear conditions and subjected to strict control.

Due to this constraint, the PCI has limited possibility to extract evidence from computer networks owned by third parties. However, like any other, the PCI may collect information from open sources, as mentioned in Section 1.3.5. The methods of the PCI are further explored later in this chapter (Section 3.6).

The PCI may gain special powers by contract, for instance, a consultancy assignment for penetration testing of the principal's computer system. If working in-house, the PCI may have powers to do things relating to the principal's computer system, which other employees may not. This will be regulated in the contract specifying the job assignment. But, in so far as the cyber investigation is directed toward computer systems other than the principal's, the PCI is not empowered to go further than any other citizen.

3.3   The Different Mandates of the LEA, NIS, and the Police

In this and the next two sections, we address cyber investigation performed by the police, LEA, and NIS, not the PCI. A cyber investigation is a *response* to high-tech or cyber-enabled crime, or to a cybersecurity incident. The response can be reactive, proactive, or preventative. To gain further understanding of the legal contexts of each of the agencies, we need to know more about the mandates that define the purpose and scope of their competence.

3.3.1   Law Enforcing Agencies and the Police

LEA will have as a principal mandate to investigate and prosecute crime. The goal is to identify the criminal and ensure that s/he is held criminally liable and is punished in furtherance of a fair trial (for further reading about the right to a fair trial, see section 3.2.3.8 in Sunde (2018)). Both the decision to bring the criminal charge and of the court to convict, hinge on the evidence. It is the purpose of the criminal investigation to uncover the facts of the crime and secure evidence of sufficient amount and quality to fulfill the burden of proof. The criminal (cyber) investigation is, therefore, a crucial part of the chain of justice. Criminal investigation and collection of evidence must be performed according to a systematic method in an impartial and unbiased manner, as explained by Nina Sunde in Chapter 2 of this book. The underlying procedural principle of objectivity is central to this process and a precondition for a fair trial (cf. page 67 in Sunde (2018)).

The main point is that a criminal investigation is typically *retroactive*, seeking as it does to shed light on a past event. Yet, the scope of the criminal substantive law (explained in section 3.3 in Sunde (2018)) influences the character of the criminal investigation. Criminal law attaches criminal liability primarily to offenses that are completed or attempted. Over the last couple of decades, however, there has been an international trend toward *proactive criminalization*, meaning that criminal liability is attached to acts taking place at earlier stages than at the stage of attempt. Conspiracy (planning and agreeing to commit a crime), and the practical preparation for it, have thus, in some instances, become criminal acts *per se*.

This is an expansion of criminal law, as such acts historically were not deemed sufficiently harmful to justify criminalization. At an early stage, the individual still has the chance to change his or her mind, reversing the plans to commit the crime. Evidence too may raise difficult issues. Manifest criminal conspiracy can be difficult to prove, except in the unlikely event that the agreement is made in writing. Several text messages pieced together, however, might be valuable in this respect. On the other hand, criminal preparation can be hard to distinguish from everyday activities.

For instance, the Norwegian terrorist who, on July 22, 2011, killed 77 people in two different attacks, first detonated a bomb placed in a car in Oslo city center. As part of the preparation, he had purchased artificial fertilizer, which he used as ingredient in the bomb. Artificial fertilizer is a perfectly legal product for use in agriculture, and the purchase was made while he lived on a farm in a rural part of Norway. According to the law, the purchase was criminal preparation of a terrorist act, but at the time of the purchase, how to prove that this was the purpose?

The idea behind proactive criminalization is that the LEA, through its investigative powers and methods, has a chance to intervene at an early stage, thus preventing the fatal terrorist act from occurring. In this sense, it is possible to talk about *preventative* criminal investigations. Yet it is retroactive, concerning as it does a crime which was committed, that is, e.g., the purchase of the artificial fertilizer with the intention to commit a terrorist act.

The LEA is not authorized to open a criminal investigation for purely preventative purposes. As already explained, the legal threshold is that there is reasonable ground to believe that a crime has been committed. An act is a crime only if this is laid down in criminal law; hence reasonable ground cannot be established at a purely preparatory stage unless such acts are criminalized. This follows from the principle of legality in substantive criminal law, see also section 3.2.3.5 in Sunde (2018). Provided the boundaries of criminal law have not been overstepped, citizens are entitled to be left in peace and not be subjected to a criminal investigation. As most crimes are not criminalized at the preparatory stage or when being agreed on, the LEA is not empowered to act against this.

Terrorism may take the form of advanced (high-tech) attacks against computer systems, though this is not all too common, possibly due to lack of spectacular effect. Terrorism can also take the shape of computer-enabled crime, such as terror financing through cryptocurrency transfers. Terrorist organizations are often associated with serious organized crime in the branches of drugs trafficking and arms trade to finance their operations and procure weapons. In practice, it may be hard to draw a clear distinction between terrorism and serious organized crime. This, in turn, has an impact on the division of powers between the LEAs and the NIS when performing cyber investigation, see Section 3.3.2.

Police organizations carry out intelligence activity typically in support of decision-making on a strategic level or in a criminal investigation. When relating to criminal (cyber) investigations, intelligence work is governed by the rules of criminal procedure that generally apply to such investigations. The need for a clear chain of command and accountability requires that the activities are gathered under a single legal umbrella, and in this context, the investigative purpose is determinative. In addition, as always, data protection rules come into play, see Section 3.2.3.

3.3.2 The National Intelligence Service (NIS)

The mandate of the NIS is to protect vital national interests and security. Relative to the cyber threat actors (CTAs) mentioned in Definition 1.3 (Section 1.2.2), the mandate typically concerns protecting vital infrastructure and central national institutions (such as government, parliament, and the judiciary), and society, against attacks from nation-states and from dangerous organized groups (non-state actors) such as terrorists. However, also "lone-wolf" terrorists may fall under the mandate. Ordinary organized crime, on the other hand, typically falls within the remit of the LEA and the police.

Nation-states set up their NIS differently but usually draw an organizational line between the part tasked with protection against internal security threats (iNIS) and external threats (eNIS). The iNIS may be an independent organization working in parallel with the intelligence function of the police and the LEA, organized under the same governmental department, usually the Department of Justice or of Internal Affairs/Home Security.

Similarly, eNIS may be an independent organization usually organized under the Ministry of Defense. However, for instance, the Netherlands has opted for partly co-organizing the iNIS and eNIS, at least with respect to their signals intelligence (SIGINT) capacity.[1] Some nation-states set up a plethora of NIS, for instance, according to its website, the US Intelligence Community consists of eighteen "elements,"[2] and (Amoore, 2020) mentions that no less than 17 US intelligence services participate in the Intelligence Community Information Technology Enterprise (ICITE) (pronounced "eye-sight") program for information processing and sharing (p. 31).

Because the mandate of the NIS concerns threats of a particularly serious nature, *direct prevention* is paramount. To this end, the nation-state will be inclined to empower the NIS to make use of privacy-invasive methods as described in Section 3.2.4. The rationale is that the consequences of a full materialization of such acts are worse than the loss of liberty caused by the NIS' use of privacy-invasive methods at a preventative stage. This rationale goes for terrorist acts, threats by state actors, and certain crimes like dealings in weapons of mass destruction.

From a legal perspective, it is important whether the mandate concerns internal or external threats. It follows from public international law that the nation-state is sovereign on its territory. The nation-state is thus free to set up the iNIS with such powers as are deemed necessary, provided the activities are controlled by the procedural guarantees and safeguards already described. Through close cooperation, the iNIS and the LEA may be able to, e.g., proactively detect terrorist activity at an early stage and prosecute.

The eNIS directs its activity toward foreign states and individuals abroad. Lacking sovereignty beyond its geographical borders, the nation-state is harder positioned when it comes to legitimately empowering the eNIS for activities abroad, see Section 3.4.1.

Uncovering the identity of the actor behind a cyber-attack or a cyber-enabled crime is often difficult (Omand, 2021). This is the problem of attribution explained in Section 1.3.3. The lack of knowledge may create uncertainty as to which organization should assume responsibility for the cyber investigation. If the adversary is a state actor, it falls within the remit of the NIS, and if an "ordinary" criminal within the remit of the LEA, unless the criminal is a "lone wolf" (terrorist). If the adversary is a combined criminal and terrorist organization, both NIS, LEA, and the police may be involved. This shows the need for close cooperation, coordination, and information sharing between the agencies to ensure clear chains of command, efficiency, accountability, and the rule of law.

3.4   Jurisdiction and International Cooperation

The lack of geographical borders makes it hard to apply rules of sovereignty in cyberspace. It is well established that a nation-state's power to perform a criminal investigation (enforcement jurisdiction) ends at its borders. Conversely, on its own territory, the nation-state's enforcement power is exclusive. A foreign investigating agency may only

---

1 The AIVD: Who we are|About AIVD|AIVD (visited November 23, 2021).
2 https://www.intelligence.gov

operate on its soil pursuant to formal consent by the host country and then normally under the lead of one of its agencies (the LEA or the iNIS).

3.4.1 The eNIS and the Principle of Sovereignty

The *Tallinn Manual 2.0 on the International Law applicable to Cyber Operations* (TM2.0) seeks to bring us a step forward in this respect. TM2.0 is developed by an international expert group in law and technology set up by North Atlantic Treaty Organization (NATO). It contains "Rules" reflecting customary international law as applied in the cyber context, see also page 4 in Schmitt (2017). The rules concern *peacetime* cyber operations and are thus relevant to our context.

The TM2.0 is not an international treaty, but for several reasons, it has gained wide international recognition. First, the expert group had members also from several non-NATO countries (i.e., China, Japan, and Kazakhstan), and the work progressed over the years with a great deal of external involvement ("The Hague process"). "Rules" were adopted only if unanimity could be achieved, and recitals provide extensive additional information and possible alternative viewpoints. Finally, the manual was reviewed by internationally renowned peers. In sum, this provides TM2.0 with strong international legitimacy.

As can be expected, TM2.0 Rule #1 states: *"The principle of State sovereignty applies in cyberspace."* This is backed by the territoriality principle underlying Rule #2 and Rule #9.

Legal Provision 3.1: TM2.0 Rule #2 on Sovereign Authority

"A State enjoys sovereign authority with regard to the cyber infrastructure, and cyber activities located within its territory, subject to its international legal obligations."

Legal Provision 3.2: TM2.0 Rule #9 on Territorial Jurisdiction

A state may exercise territorial jurisdiction over:

a) cyber infrastructure and persons engaged in cyber activities on its territory;
b) cyber activities originating in, or completed on, its territory; or
c) cyber activities having a substantial effect on its territory.

The manual explains that "cyber infrastructure" and "cyber activities" means the *"physical, logical, and social layers"* of computer systems and networks on the country's territory (recital 4 to Rule #1) (Schmitt, 2017). The reference to "territory" implies that the manual, unfortunately, does not really guide the cyber investigator as to what to do when the whereabouts in cyberspace are geographically uncertain or unknown.

Foreign intelligence (eNIS) has been carried out at the order of the sovereign throughout history (Stenslie, 2021). The activity is not governed by international conventions, and information about cyber operations is not made public.

For several decades now, there has been a noticeable development in nation-states' use of bulk collection of Internet and communication data, a fact which first became prominently public by the Edward Snowden revelations (Greenwald, 2014). However, as bulk collection became publicly known, governments were forced to strengthen both the rule of law and democratic control with the foreign intelligence service. It is conceded that it is not possible to limit such methods only to concern data relating to individuals located beyond national borders, so special protective measures must be implemented to protect individuals residing on the territory from being subjected to surveillance.

According to democratic principles and the rule of law, the methods ought to be described and laid down in law, and mechanisms established that protect the national population against interference from eNIS. Liberal democracies use systems of judicial review and independent oversight to keep the activity under control. The control, however, principally concerns the protection of the rights of individuals residing in the nation-state. There is less concern for the rights (and fates) of foreigners living abroad.

3.4.2 The iNIS and the LEA – International Cooperation

As TM2.0 applies to cyber operations in peacetime, the rules are also relevant to cyber investigations performed by the iNIS and the LEA. Clearly, the limits of territorial enforcement jurisdiction apply to their activities as well. However, several arrangements and initiatives on national and international levels are taken to accommodate the needs of the respective agencies in uncertain circumstances online.

Firstly, there are procedures for international assistance in procuring and sharing (digital) evidence in a criminal investigation. Such regimes for mutual legal assistance are described in section 3.5 in Sunde (2018).

Secondly, in Europe, both the European Union (EU) and the Council of Europe have for years been working on international procedures regarding e-evidence. The aim is to ensure that an LEA in nation-state A can directly order a service provider in nation-state B to produce data generated from its services. Furthermore, providers registered in a third country, offering services to European citizens, shall have a registered point of contact in Europe able to produce the data lawfully requested by the LEA/iNIS pursuant to a production order (production order is explained in section 3.4.5 in Sunde (2018)). Until in place, the more time-consuming procedure of mutual legal assistance applies.

Thirdly, setting up joint investigation teams (JIT) with LEA members from different countries under the procedure provided by Europol and Eurojust has proved effective. This possibility is open not only to EU member states but also to third countries. The JIT concept is described in section 3.5 in Sunde (2018).

On a national level, there is a trend toward legislating powers to unilaterally secure evidence located on servers abroad (or at an unknown location). Typically, this will entail methods such as the use of "equipment interference" (United Kingdom) or "network investigation technique" (United States). These are references to secret access to and monitoring of computer systems and networks. Legislation of this kind clarifies to the cyber investigators what s/he is empowered to do as seen from the national jurisdiction's point of view. It further clarifies to the outside world the actions the said jurisdiction, under certain circumstances, feels entitled to take toward servers and systems abroad.

However, it does not establish a *right* as per public international law to do this against servers, networks, or closed social media groups, etc., (cf. *"physical, logical, and social layer"*) located in a third country. Depending on the circumstances, the target country might claim a violation of its sovereignty. Yet, it cannot be taken for granted that such cyber investigation methods *by default* interfere with principles of sovereignty and territoriality. Public international law is said *not to concern itself with trifles (de minimis non curat lex)*. Applied to the cyber context, it means that for the activity to be an interference, it must amount to a certain noticeable disturbance, i.e. exceed a certain threshold (Skjold, 2019). This is supported by TM 2.0 Rule #9 (c), according to which a country only assumes jurisdiction if the cyber activity has "a substantial effect" in its territory (Rule #9 is cited in Section 3.4.1).

For instance, the use of login data to accede a user account on Amazon Web Services was not deemed to exceed this threshold by the Norwegian Court of Justice, in a case where login data was obtained from the account holder who was also present when the cyber investigator accessed the account (HR-2019-610-A). Nation-states take different views on these questions, so there is no international consensus yet. It is interesting that in Scandinavia, a region characterized by shared values and constant cross-border interaction, Denmark and Norway accept this method. In contrast, the third neighbor, Sweden, does not (as per the time of writing).

3.5   Human Rights in the Context of Cyber Investigations

TM2.0 Rule #34 states, *"International human rights law is applicable to cyber-related activities."* The rule reflects that the law applies equally to life in cyberspace and physical space. In other words, cyberspace is legally considered an integral part of life in general. See section 3.2.3 in Sunde (2018) for an overview of relevant legal human rights instruments.

The manual continues stating:

Legal Provision 3.3: TM2.0 Rule #36 on Human Rights

"With respect to cyber activities, a State must (a) respect the international human rights of individuals; and (b) protect the human rights of individuals from abuse by third parties."

3.5.1   The Right to Fair Trial

Criminal investigations by LEAs are ordinarily "open," entailing a right for the suspect to be informed *"promptly"* of the criminal charge against him/her and a right to be represented by a defense lawyer. The suspect further has a right to be presumed innocent until proved guilty according to law, and police interviews shall be voluntary with the possibility to have the defense lawyer present. The suspect is also entitled to be present while the criminal investigator collects evidence located in his or her private sphere, for instance, search or seizure of electronic devices. These rights are embedded in *fair trial* (also known as "due process"), cf. the ECHR article 6, and provide the suspect with an opportunity to rebut the criminal charge and challenge the steps taken against him/her. The rules apply equally to cyber as to other criminal investigations. Fair trial is covered more extensively in section 3.2.3.8 in Sunde (2018).

However, identifying the perpetrator can be a challenge (cf. the attribution problem described in Section 1.3.3), and s/he cannot be informed of the criminal charge when the identity is unknown. In this event, the investigation is categorized as being performed on a "case" level instead of on a "personal" level, and in principle, it is open.

We now turn to "covert" investigation, which is practical in a cyber context.

3.5.2 Covert Cyber Investigation

By "covert criminal investigation," we mean "a criminal investigation performed with use of intrusive investigation measures, without informing the suspect even if the identity is known."

Definition 3.1: Covert criminal investigation

A criminal investigation performed with use of intrusive investigation measures, without informing the suspect even if the identity is known.

Covert privacy-intrusive investigation methods are problematic in democratic and human rights perspectives. Since the suspect is not aware of what is going on, s/he cannot rebut the charge or explain relevant matters to the police. Covert methods, therefore, carry a greater risk of abuse, and the suspect may over time suffer severe intrusions of the private sphere. The methods may obviously affect family members and the suspect's social network as well, and not least, the fundamental right to have confidential contact with one's lawyer.

It is fundamental that the LEA, iNIS, and the police are prohibited from obtaining and possessing communication between lawyer and client, whether in electronic or physical form. If this still occurs, for instance resulting from bulk collection of communication data or the imaging of a hard drive, the privileged material must be removed immediately from the agency's possession, or at least – as a minimum – be made inaccessible in a bulletproof way. The ECtHR zealously guards and reinforces this protection (see in the context of search and seizure chapter 12.4 VII in Harris (2018) and chapter 16.7 in Kjølbro (2017)). For an analysis specifically relating to privileged material in digital evidence, see chapters 16 and 17 in Sunde (2021b). See also Kozhuharova (2021).

The government with its agencies – here: the LEA and iNIS – is a dangerous guardian and must be kept at bay through clear laws, judicial review, and independent oversight. This was explained in Section 3.2 and is here reconfirmed with respect to cyber investigations by reference to TM2.0 #Rule 37. See also section 3.2.2.6 in Sunde (2018).

Legal Provision 3.4: TM2.0 Rule #37 on International Human Rights

"The obligations to respect and protect international human rights [. . .] remain subject to certain limitations that are necessary to achieve a legitimate purpose, non-discriminatory *(sic!)*, and authorized by law."

The use of covert investigation methods must not be kept secret indefinitely. The main characteristic of a covert method is that the suspect is not notified at the time when it is carried out. Usually, the suspect has a right to be informed afterward, at least before trial. Pursuant to the procedural principle of *equality of arms*, the defendant has a right to access the complete case file documenting the investigation, and all the evidence, in advance of trial in order to prepare the defense. There are exceptions, though; when vital national interests are at stake, the period of secrecy may be extensive, sometimes lasting beyond the suspect's/convict's lifetime. But then special procedural guarantees are in place to compensate for the secrecy.

In line with Lentz (2019), we distinguish between covert methods by technical means ("technical hacking") and by use of social skills ("social hacking").

3.5.3 Technical Investigation Methods (Technical Hacking)

With "technical hacking" we think, for instance, of:

— Gaining access to a protected computer system or network
  ○ Listen in, monitor or copy electronic communication/network traffic
  ○ Perform digital search and seizure
  ○ Monitor the activity on a computer system, for instance, to circumvent encryption, obtain login data, or gain access to hidden services
— Gathering metadata or geolocation data from a computer system or network.

The methods are aimed at gathering information from digital sources. However, hacking may also make it possible to come in position to activate audio and video functions on the device, with the ensuing possibility to monitor physical activity. For example, seeking to catch the communication in a meeting between members of a criminal group, the LEA remotely activates the microphone of one of the smartphones in the room. The method is particularly intrusive, especially when applied in private homes where it provides the police with a covert *de facto* presence in the inner part of the private sphere. Some jurisdictions do therefore not authorize it, or only to a limited extent. For instance, Norwegian criminal procedural law authorizes covert audio surveillance in private homes, but not covert video surveillance.

Covertness is an inherent feature of the methods on the list. The exception is search and seizure, which frequently also take place with the suspect's knowledge, for instance, during a house search. The Cybercrime Convention article 19 regulates computer search and seizure. The provision does not say whether the method may be performed covertly, which is for the nation-state to decide. This is often permitted in relation to investigations concerning serious crime. For more on this, see section 3.4.4 in Sunde (2018).

The methods characterized by covertness are partly regulated in the Cybercrime Convention article 20 (real-time collection of traffic data) and 21 (interception of content data). LEAs worldwide, however, are eager to exploit innovative use of the technology, and the convention has not kept up with the developments in investigative methods.

First, we address the *monitoring of computer activity* (also known as "computer surveillance," "equipment interference," or "network investigation technique"). The method does not have a separate provision in the convention, but many jurisdictions authorize it by special provisions in their criminal procedural law. The method is aimed at gaining access to protected resources. However, as shown by Lentz (2019) it is not always clear how far one may go in this respect. The assessment largely hinges on concrete circumstances. Indirectly the method's gray zones illustrate the problems with establishing a clear legal concept of illegal access as a criminal offense (as per the Cybercrime Convention article 2, cf. section 3.3.3.1 in Sunde (2018)). In any case, the method is controversial because of the risk of causing damage or unintended effects on the target system, including exposure of data belonging to third parties, e.g. thousands of account owners on an email server.

Second, we address *international operations* concerning interception of electronic communication such as Venetic, led by French LEA, and Trojan Shield, led by the Federal Bureau of Investigation (FBI) in the United States and Australian LEA. The LEAs gained secret control over encrypted communication services mainly used by criminals. In Operation Venetic, the French police hacked the communication network EncroChat and covertly established itself as a service provider. Thus, all communication was intercepted. After EncroChat was taken down, criminals experienced an abrupt need for a new secure communication service. This was provided by the FBI in cooperation with Australian LEA, which through intermediaries uniquely targeting criminal clients, offered the encrypted communication service Anon. In both operations, hundreds of suspects were apprehended all over the world, charged with crimes such as trafficking in weapons and drugs, murder for hire, kidnapping, and armed robbery.

Unsurprisingly such operations raise numerous legal issues (Sunde, 2021c). Secret surveillance of private communication must be suspicion-based and relate to specific suspects. This condition has traditionally been understood to require that the suspicion is based on concrete circumstances. Secret surveillance must, therefore, primarily target the suspect's device. Sometimes it can also be extended to related persons, yet it must still be individually targeted subject to very strict conditions. Venetic and Trojan Shield, on the other hand, collected communication from hundreds of people, perhaps thousands. So far, there has not been public information available confirming that there was a concretized suspicion against each one of those when the operation was initiated. It indicates that the coercive measure was applied to establish a suspicion that, as per the law, should have existed beforehand.

Further, news media report that EncroChat was also used by noncriminals who felt a need for a secure communication alternative. Their exposure to the LEA is collateral, the extent of it impacting the proportionality of the operation. It is likely that Trojan Shield was more targeted, as the LEA tried to avoid distributing Anon to others than known criminals.

The duration of such operations raises serious dilemmas for the LEAs. Trojan Shield was going on for months, perhaps a couple of years, and during this time, the FBI was said to have become aware of numerous serious crimes, including kidnapping, extortion, armed robbery, and murder. The question is, for how long can an LEA be a passive bystander in search of "the even bigger fish"? The LEA has a duty to protect people from crime and actively intervene when needed to protect life and health of potential victims. Is a criminal's life worth less than that of a noncriminal? If not careful, the LEA even risks criminal liability for collusion.

There are jurisdiction issues and forum shopping as well. As explained in Section 3.4.1, the legal power to perform a criminal investigation ends at the geographical border. EncroChat and Anon were used by individuals (criminals) internationally. French LEA has asserted it hacked EncroChat's server located in France, hence was subject to French enforcement jurisdiction. Contrary to this, news media reporting from open court proceedings against the suspects maintain that the interception took place at the user endpoints. If so, French police may have exceeded the territorial limitation of enforcement jurisdiction, as users resided in many countries other than France. This could have adverse effects on the admissibility of the evidence in the criminal proceedings, but it is also possible that any weaknesses are mitigated by Europol's and Eurojust's involvement in the operation. With regards to Trojan Shield, the manner of distribution of the Anon handsets may be questioned. If they were distributed to criminals residing abroad without permission from the LEA in the affected country, also this raises a jurisdictional issue.

Venetic and Trojan Shield concern hacking of communication services. Comparable operations have been carried out online in order to take control over criminal marketplaces. This has been seen, e.g., in relation to Alphabay I and II and Hansa (Dordal, 2018). A difference though, is that, on the dark web, one hardly knows the location of the market. Hence the significance of jurisdictional issues is downplayed.

The dilemmas we have described are relevant to cyber investigations. The baseline is that the method must be lawful and duly authorized and controlled, as per the law in the country in charge of the investigation. In other words, that Venetic was lawful in relation to French law, and Trojan Shield in relation to the US and Australian law. This requirement is necessary to ensure that such invasive methods are not performed outside the law (democratic control and the rule of law) without judicial control and independent oversight. It is also important that such operations are well planned and documented, the purpose clearly defined both to know what one seeks to achieve and when to stop.

The operations also illustrate a need for legal conceptual development relating to coercive measures, as many jurisdictions today struggle in the application of traditional concepts to new technological possibilities, innovative user behavior, and the challenge of encryption.

The Cybercrime Convention article 15 reminds the legislator that the legal framework of privacy-invasive methods must respect the conditions and safeguards embedded in human rights law. Correspondingly, TM2.0 Rule #35 states that individuals *"enjoy the same international human rights with respect to cyber-activities that they otherwise enjoy."*

Until now, we have addressed a context where the LEA seeks to obtain the digital evidence directly by its own efforts. The law provides an alternative in the form of a production order (also mentioned in Section 3.4.2). By this legal instrument, the LEA can order a third party in possession of the data to hand them out to the LEA. Production order is commonly used to obtain metadata/electronic traces in service providers' possession. It may also be applied to obtain content data, e.g. from cloud providers, but in that case, the procedure is usually intricate and time-consuming. A similar procedure is available for performing covert communication surveillance with the assistance of the communication service provider.

In addition, the Cybercrime Convention article 19(4) authorizes the criminal investigator to order *"any person who has knowledge about the functioning of the computer system . . .to enable the undertaking of [computer search and seizure]."* Thus, the cyber investigator may, in the context of a criminal investigation, for instance, order IT personnel to provide access to a corporation's computer system.

3.5.4 Methods Based on Social Skills (Social Hacking)

Investigation methods such as covert observation and infiltration make use of social skills. The cyber investigator can go online with a user profile not disclosing the LEA affiliation or put a police robot (typically a chatbot) pretending to be a real person, to work online. To ensure accountability, European data protection rules and AI principles require a human in the loop. Hence someone in the LEA must operate or otherwise oversee the actions of the robot.

By "infiltration," we mean "a method whereby the cyber investigator gains information through interaction with others while concealing his or her professional identity."

Definition 3.2: Infiltration

A method whereby the cyber investigator gains information through interaction with others while concealing his or her professional identity.

For instance, aiming to find out whether a virtual marketplace offers illegal goods, the investigator may pretend to be a potential buyer and ask for offers. Another example is a chatbot pretending to be a minor to get in contact with potential child sexual offenders, who can be tracked down by the cyber investigator. The method is known, i.a., from the Sweetie-project, and (Sunde & Sunde, 2021) describe how the technology can be further developed by the use of AI and text-based analysis. Legal implications of the method are analyzed, i.e., by Schermer *et al.* (2019) and Sunde and Sunde (2022).

"Observation" is "covert presence over time to systematically obtain information about the suspect and his or her activities."

Definition 3.3: Observation

Covert presence over time to systematically obtain information about the suspect and his or her activities.

As an example, let us assume that the cyber investigator is present at an illegal online marketplace in order to gain knowledge about the activity. This is observation. Perhaps it was first necessary to manipulate somebody to gain access to the marketplace. This is infiltration. In practice, observation and infiltration often go hand in hand and may be hard to distinguish.

As part of the right to privacy, a person is free to decide with whom to relate and socialize. Gaining access to the suspect's private sphere under false pretenses is, therefore, a serious interference of privacy. For this reason, the methods must be legally authorized, necessary and proportionate, as already explained.

In her ground-breaking thesis (Lentz, 2019) identifies important gray zones in online infiltration. This is primarily due to the semi-public character of many such fora. How is one to assess the status of a group established for the purpose of arranging a large family gathering, perhaps consisting of two hundred persons? Does it matter if the whole party belongs to the Cosa Nostra, or only a few do? Is such a group "private" at all, considering the large number of members?

We now turn to *fair trial* implications of infiltration. The right to fair trial restricts the use of police provocation (entrapment) and bars evidence obtained in violation of the right to protection against self-incrimination. Both are pitfalls to infiltration.

Entrapment: A cyber investigator purchases stolen credit cards and malware and rents a botnet under the pretext of wanting to launch a distributed denial-of-service (DDoS) attack. Objectively, the seller contravenes criminal provisions prohibiting information fencing (the stolen credit card numbers) and the making available of malware and services designed for attacking computer systems, cf. the Cybercrime Convention article 6 (see section 3.3.3.3 in Sunde (2018)).

Legal doctrine says that the LEA should not cause crimes that otherwise would not be committed. Hence, because the criminal act was prompted by the criminal investigator's request, criminal charges against the seller shall not be brought.

Yet, provocation may be used to gain evidence for *ongoing* crime, such as on online marketplaces where illegal trading is regular business. The rationale is that the crime would have been committed in any case: had the "customer" (the cyber investigator) not turned up, another real customer would. The cyber investigator's act only changed the time for a crime that nonetheless would have been committed. The alteration is not material; hence the crime can be prosecuted.

Provocation may be used to procure evidence for a crime that was *already committed*, such as illegal possession of malware or child sexual abuse material. These are criminal offenses, as per articles 6 and 9 of the Cybercrime Convention (Sunde, 2018, ss. pp. 88 and 91–92). If the cyber investigator succeeds in obtaining illegal material, the material is evidence for prior unlawful possession, and for this, the suspect may be prosecuted. Charges cannot be brought, however, for the *sale* to the cyber investigator, as this was prompted by the investigator's request unless, of course, such trading was regular business.

Use of a LEA chatbot in a public chat room pretending to be "Lisa 12," must consider the prohibition against unlawful provocation. If Lisa is passive, the initiative is on the would-be offender. Regularly the suspect may be held criminally liable in this situation. With today's knowledge about "sextortion," one should consider charging for attempted rape, not only for lascivious speech or attempt to procure sexualized images of a child. The chatlogs documenting the interaction can be produced as evidence at trial (Sunde & Sunde, 2022a).

The presumption of innocence entails a right not to incriminate oneself, and a suspect is free to decide whether to accept being interviewed by a criminal investigator. Depending on the circumstances, a chat conversation between a cyber investigator ("Lisa 12") and a suspect can amount to a covert police interview. If so, the suspect is deprived of his or her right to be silent and be assisted by a defense lawyer, and this is irreconcilable with the right to fair trial. Covert police interviews are, therefore, as a rule, not accepted as evidence at trial, and the right may, of course, not be circumvented by the cyber investigator using a third party – such as a police robot – to engage in the conversation.

For a conversation to be regarded as a police interview, the criminal investigator must have taken the initiative and/or been active with follow-up questions leading the conversation in a certain direction. The passive presence of "Lisa 12" and the ensuing conversation on the initiative of the would-be offender is therefore not necessarily problematic in relation to fair trial.

Information that is inadmissible as evidence might still be used as *a lead* in a further criminal investigation. The purpose is then, in the next phase, to secure evidence that may be produced at trial, for instance, by traditional investigation methods in an investigation that is "open." For instance, "Lisa 12" may have been too active online hence the chat conversation cannot be used as evidence. The conversation has, however, established reasonable ground to suspect the conversation partner of being a serial offender. The electronic traces collected by Lisa in the chat are valuable leads to trace back the location and identity of the suspect. The purpose in this stage of the criminal investigation would be to uncover facts relating to those sexual offenses (not relating to Lisa), committed against other children.

Assessment of the legality of police provocation and online conversations is highly contextual, depending on concrete circumstances. Important checkpoints are, whether the police behaved in an essentially passive manner; the operation was authorized by a competent person; was well planned with a clear goal as to what to be achieved; and, whether the documentation is clear, reliable, and complete, made close in time to the event, thus enabling contradiction by the defense lawyer and judicial review.

Finally, the law takes a more lenient view when the cyber investigator has obtained consent from a person who is party to the conversation. This could be relevant, for instance, when gaining access to a shared email account used by criminals as a message board. If the cyber investigator posts and reads messages by consent from one of the users, this could be acceptable in relation to the protection against self-incrimination. Another example is participation in a closed group on social media, with consent from one of the participants.

In the end, the police operations mentioned in this section must be assessed independently on their merits. As national procedural law varies, it is hard to come up with definite, clear rules.

3.5.5 Open-Source Intelligence/Investigation

Information collection from open sources is known as OSINT. Social media intelligence (SOCMINT) is related but not entirely identical. In this case, information is gleaned from social media platforms, which by implication makes privacy and data protection rights an issue. Although such fora may be accessible to many people, they have not consented on a general basis to the use of personal information for a different purpose than that of the forum.

Everyone is free to collect information from open sources on the Internet. With respect to LEAs, this right is made explicit in article 32 (a) of the Cybercrime Convention (see section 3.4.2.1 in Sunde (2018)) but is true also for the police, the NIS, and private individuals such as the PCI.

The hard question is, what exactly is an "open" source? Note that Definition 1.12: Open-Source Intelligence in Section 1.3.5 *"publicly available information that has been discovered, determined to be of intelligence value, and disseminated by an intelligence function"* does not give the answer. Definition 1.12 concerns information after it has been processed by the intelligence agency, whereas here, we are focusing on *the source* from which the information (to be processed) is collected.

There is no clear answer to the question. It seems uncontroversial that information behind a paywall is free, provided anyone willing to pay can get access. Neither do access restrictions based on geo-localization techniques affect the openness of the source. Circumvention of such restrictions might give rise to a dispute with the content provider but is not relevant to the question of openness in the present context. Hence, sources of this kind are open.

In relation to SOCMINT (Rønn, 2022) notes that many NIS and LEAs make use of *"sophisticated software tools designed to penetrate otherwise closed social media platforms and/or by exploiting the ignorance of Internet users in terms of privacy configurations on Facebook or other social media platforms"* (p. 159). What Rønn describes is a forum that is not freely accessible, not even against payment. In such a case, the source may hardly be regarded as "open." Rønn further concludes that *"open"-source intelligence seems to be situated in a grey zone between private and public information, and the exploitation of such information raises new questions about the nature and privacy of online information"* (p. 159).

Clearly, data protection rules come into play, i.e. the Police Directive and the GDPR. Importantly, data protection rules only regulate data processing *after* the data are collected. The rules do not give a legal basis for the data *collection*. The right to collect personal data needs a separate legal basis. Either the data are free because they are in the "open," or the collection needs basis in other legal rules because the collection would interfere with privacy rights.

3.6 The Private Cyber Investigator

In Section 3.2.2, it was explained that the PCI must comply with criminal law, the contractual relationship with the principal, data protection law, and employees' privacy rights. In this section, we go into this in more detail.

3.6.1 Cyber Reconnaissance Targeting a Third Party

Criminal law provisions protecting cyber security draw a line for how far the PCI may go in a cyber investigation that targets the computer system or network of a third party. Broadly such provisions protect against illegal access (including monitoring) and cyber vandalism. However, the Cybercrime Convention does not impose an obligation to criminalize *cyber reconnaissance*, and the question is how far the PCI may go in this activity without trespassing the law.

Criminal law prevents the PCI from penetrating the system perimeter, listening in/ recording/copying network traffic (illegal access), and performing activities that interfere with the normal operation and functioning of the system (cyber vandalism). As a rule, the acts must be performed *intentionally* (see pages 77–78 in Sunde (2018)), yet in some instances, the law attaches criminal liability to *negligent* acts as well.

In addition, criminal law makes *attempt* a punishable offense. This follows from traditional doctrine and from the Cybercrime Convention article 12(2), specifically in relation to illegal access and cyber vandalism. For an attempt to be criminal, it must be committed with intent to complete the crime.

The purpose of cyber reconnaissance may be to learn as much about a computer system or network, and its owners/users, as possible. Assuming that the aim of the PCI is to glean information from exposed parts of the computer system *and not go any further*, the activity may not be regarded as a criminal attempt to obtain illegal access because intent to complete such a crime is lacking.

However, an activity that is too intensive, persisting, and prying into protected resources may be provocative and perceived as an attempt to gain illegal access. The question of intent hinges on the evidence, and the more prying, the less convincing a claim that gain of access was not intended. Moreover, vulnerabilities in the system may not be exploited, as this would amount to regular hacking, see section 3.3.3.1, page 83, in Sunde (2018). So, if the PCI understands that the "reconnaissance" exploits a resource unintentionally exposed by the owner, the boundary of the law might be overstepped.

Reconnaissance is ideally performed without triggering interest from the targeted party or bystanders. In the cyber context, "bystander" could, for instance, be a network security provider. However, in relation to the risk of negligently becoming liable for cyber vandalism, it is relevant to consider *unintended* effects on the target system caused by the activity. In short, if it interferes with functions of the system directly or indirectly, or noticeably consumes system capacity, the activity could be deemed as negligent vandalism.

Note that although a nation-state is not obliged by international criminal law to criminalize cyber reconnaissance, it may have chosen to do so. Furthermore, criminal law normally strikes down on industrial espionage, and cyber reconnaissance might amount to this (depending on the circumstances). To be on the safe side, the PCI should check the criminal law of the jurisdiction where the target system is located.

3.6.2 Data Protection and Privacy Rights

The information collected and analyzed by the PCI is to a large extent, personal data, hence in Europe, the PCI must comply with the rules of the GDPR. The GDPR is an extensive and detailed set of rules with direct effect in the EU/EEC (European Economic Community) member states, and for a professional PCI, it is essential to know these rules.

Finally, the PCI may suspect that persons working for the principal (system/network owner) are a security threat. It could be for a personal purpose, for a purpose shared with an adversary of the principal, or because one unwittingly is manipulated by an adversary. The PCI feels a need to secretly monitor the employee's digital resources to clarify the circumstances, but would this be lawful? Firstly, the law recognizes a right to privacy related to electronic communication also at the workplace, as discussed in Halford v UK (1997) and Copland v UK (2007). The principal, however, obviously has legitimate interests in the security of the digital resources in use in the business operation (Barbulescu v Romania, 2017).

For the PCI to go to such steps, what is needed is at least that the corporation (principal) has implemented clear internal guidelines regulating the circumstances of such surveillance, the procedure to be followed, the involvement of a trusted representative of the employees (and the union if applicable), and provides clear and reliable documentation of any such surveillance. The general procedure must be communicated to the employees.

3.7 The Way Ahead

The complexity and high level of detail of data protection rules may curb the freedom needed in a criminal (cyber) investigation. Given the ever-increasing importance of data protection rules, one may finally come to a point where the aims and principles of criminal procedural law and data protection law clash. This tension might become a driver for disputes to be solved by the court and by the legislator. Further, the tension might exacerbate when AI makes its way as a regular component of cyber investigations. To a large extent, European principles of AI regulation build on data protection principles, and the proposed Artificial Intelligence Act heralds a strict regime on the use of AI for law enforcement, as opposed to commercial purposes.[3] However, increased use of AI in cyber investigations is a trend which must be expected to continue and grow stronger.

A common theme in *Intelligence Analysis in the Digital Age* (Stenslie, 2021) is that information scarcity is largely a past phenomenon. Nowadays, information overload is the problem. From a legal perspective, this raises increasing concern relating to the proportionality condition, which is integral to privacy rights and data protection law. The effect is a demand for increased efforts to keep each step of the cyber investigation very targeted and to minimize the data volumes as much as possible, both initially and under the course of the investigation.

3.8 Summary

In this chapter, we have reviewed legal norms relevant to cyber investigations, covering both criminal investigations performed by the LEA and other cyber investigations performed by the NIS and the PCI. The chapter has reviewed the limits of investigative freedom within the frameworks of public international law, human rights and data protection law, and criminal law. Finally, the legal mechanisms for international cooperation in cyber investigations and international evidence exchange have been reviewed.

3.9 Exercises

1 Explain the mandate of the LEA and the purpose of a cyber investigation performed by the LEA.

2 Explain the mandate of the NIS and the principal purpose of a cyber investigation performed by the NIS.

3 Explain to whom the LEA and NIS are accountable.

4 Explain the general conditions for lawful application of privacy-invasive cyber investigation methods.

5 Explain the control mechanisms seeking to ensure LEA/NIS accountability.

---

3 Proposal for a regulation of the European Parliament and of the Council laying down harmonized rules on artificial intelligence (artificial intelligence act) and amending certain union legislative acts. Com/2021/206 final. https://eur-lex.europa.eu/legal-content/EN/TXT/?uri=CELEX:52021PC0206.

6   Explain the legal framework of the PCI.

7   Explain why privacy-intrusive cyber investigation methods are unlawful to the PCI, yet may be lawful to the LEA/NIS.

8   Explain legal reasons why international cooperation may be necessary for a cyber investigation.

9   Explain the legal limits of cyber reconnaissance.

10   Explain what "open source" on the Internet is and the concerns causing "gray zones" relating to information collection from such sources.

4

Perspectives of Internet and Cryptocurrency Investigations

Petter Christian Bjelland*

Oslo, Norway

Modern investigations usually involve data accessible through the Internet or data generated from Internet activity. This data can be crucial to answering the required questions, from determining who uses a particular Internet Protocol (IP) address at a particular time to analyzing Internet usage on an acquired device. This chapter will look at typical sources of information on the Internet, how to access them, and how to analyze them. We start with a couple of plausible examples based on real cases before diving into four different sources of information related to the Internet or its use: networking, open sources, web browsers, and cryptocurrencies. We will then look into how collected data can be prepared and analyzed.

The acquired data may not be in a readily available format, and thus some techniques for extracting relevant information from data formats will be presented. Further, some analysis techniques related to link diagrams (graphs), timelines, and maps will be presented with examples before discussing how these analyses can be presented in reports.

4.1 Introduction

Knowledge of how the Internet is glued together is fundamental to interpreting technical data we collect but is also central to being creative about what data may be available and how to make use of it. This chapter is based on our previous introduction to the Internet Forensics (Bjelland, 2018) in the Digital Forensics textbook (Årnes, 2018), which we recommend reviewing to provide the necessary foundation for this chapter. In this chapter, we extend this work with techniques for how this information can be traced and acquired and how it can be interpreted. First, we will look at some example investigations that included Internet and network data.

---

* Petter C. Bjelland holds an MSc from NTNU and has previously worked with the Norwegian Criminal Investigation Service (Kripos). He has international experience conducting data processing in large enterprise investigations.

Cyber Investigations: A Research Based Introduction for Advanced Studies, First Edition. Edited by André Årnes.
© 2023 John Wiley & Sons Ltd. Published 2023 by John Wiley & Sons Ltd.

4.2   Case Examples

The goal of presenting the case examples is to help put the theory into context. These examples are all based on actual investigations that the author has been involved in but have been made notional. We start with an Internet salesperson that forgot to cover their tracks – getting an idea of what the perpetrators are attempting to achieve and why may be a critical part of identifying which traces may have been left behind.

4.2.1   The Proxy Seller

This first example looks at a person selling software online to help customers connect to hacked computers in an untraceable way. Clues left behind within the promoted application binary provide clues into the seller's identity.

Example 4.1: The proxy seller

This first example looks at a person who was selling software online designed to help customers connect to hacked computers in an untraceable way. Traces left behind within the promoted application binary provided clues into the identity of the seller.

The proxy application showed up on the author's radar as one of the promoted IP addresses was claimed to be in Norway. Looking up the application name using a search engine, the author identified several different sites where the application was being promoted. Several different user and display names were used for this, and there seemed to be no clear pattern. Surprisingly, the download link from the advertisements allowed the download of the application with no additional questions asked.

The analysis of the proxy access application resulted in a probable username connected to the promoted proxy software, raising the question of whether this person also used this username services out on the Internet?

One display name appeared connected to the username on multiple services. The name was common enough that a search on Google for it returned a lot of results, but one of the top ones was from a known video sharing service. The videos the user had uploaded were related to a completely unrelated subject. There was, however, one video that looked different. In this video, presumed addresses of the user, as well as other identifiable information, could be observed.

4.2.1.1   Background

How packets are transmitted across the Internet means all data eventually can be traced back to the originating computer. Even when, or rather because of, using technology developed to obfuscate traces on the Internet, police could identify the sender as a supposedly anonymous bomb threat as he was the only one using this particular technology at the time (Brandom, 2013).

Obfuscating or otherwise masquerading the true originator on the Internet is often called proxying. Many uses of proxies are entirely benign. For example, your employer may require you to log into the office network when working remotely. Other use cases are more in the gray area. For example, when your favorite television (TV) show is not available in the country you are in, you can pay a provider to let you proxy your data through their servers, which gives you access anyway. And then we have the outright malicious: Cases where someone breaks into other people's computers to send data through their machines without their knowledge. This latter kind is an enabler for different types of cybercrime, such as spoofing physical locations.

The developer created an application that can download a list of "hacked" login details associated with IP addresses and lets the user select which country to proxy through and connect with a click. The seller had created a platform for trading between those who compromise and gain access to people's computers and those who would like to use this access for other purposes. The list of available IP addresses and login information can be purchased from hackers and passed on to customers. This login information sometimes points to servers that can be relied upon to stay available. Still, often it points to average Joe's 10-year-old laptop that runs a fantastically outdated operating system.

The application is built with a graphical user interface that makes it trivial for a user to simply choose the desired country to connect to and then automatically connect to one of the recently provided IP addresses. As an added service to customers, connections are also passed through an onion circuit using The Onion Router (TOR) to hide their traces. The application is promoted on several different sites on the Internet, but it is not clear how successful it turned out to be.

The proxy application showed up on the author's radar as one of the promoted IP addresses was claimed to be in Norway. Looking up the application name using a search engine, the author identified several different sites where the application was being promoted. Several user and display names were used for this, and there seemed to be no clear pattern. Surprisingly, the download link from the advertisements allowed the application to download with no additional questions asked.

4.2.1.2 The Proxy Access Application

There are different ways code written by humans is converted into instructions understood by computers. In this case example, we briefly describe three different categories: assembly code, scripts, and virtual machines. Assembly code is a collection of bytes arranged in an order a computer can evaluate and run directly.

The given CPU architecture knows where, in the collection of bytes ("the binary"), the first instruction is located and how to read each instruction ("if this then do this, otherwise do this"). Because instructions are specific to each CPU architecture, code written by humans must be converted ("compiled") to one collection of bytes for each architecture the program should run on. This compilation requirement led to the creation of virtual machines, where instead of compiling code to binaries, it is converted into a format that another program ("the virtual machine") understands.

This program is, in turn, compiled for each architecture the written code should run on. This lets the writer of the code without too much consideration about the system it will run on, and instead, let the virtual machine take care of compatibility. The cost is less control over how specific code instructions are performed. Java is one popular virtual machine. Finally, scripts are human-written code read and converted into CPU instructions as the program runs. Python and Javascript are popular scripting languages.

The proxy seller was using a virtual machine language to develop the product. Building programs using virtual machines often come with a build system. This system is responsible for converting the written code into instructions the virtual machine understands. These systems can be rather complicated, and it' is not always easy to know exactly how the converted package is created. In this case, the build system had included information about where on the coder's computer system the source code was located; this included the presumed username of the currently logged in user, like:

C:\Users\proxyseller-123\super-proxy-v1

The output of generated virtual machine code often takes the form of a single-file archive, such as a zip file. It' is therefore not readily visible to anyone what information the build system may have put into it, but there are a couple of ways we could look for it.

Decompilation is the process of going from a collection of bytes representing instructions for a specific CPU architecture to a representation that humans can more easily understand. It is an essential part of areas related to computer investigations, like malware research (understanding behavior) and security audits (finding flaws that can be exploited). The process is often referred to as reverse engineering, and IDA Pro ("the Interactive Disassembler") and Ghidra are two popular tools that enable it. One of the standard features of these tools is the extraction of static string variables, which would, in this case, make it possible to search for the above folder path, for example search for "users."

Instructions created for virtual machines can similarly be converted into something readable for humans. In the case of Java, it is possible to retrieve something reasonably identical to the source code. As the build system added the folder path, we would not find it within the compiled instructions but rather in metadata files added to the same package. If we unpack the downloaded application, we can browse and open the included metadata files and eventually find the folder path.

There is a simpler way still to identify possibly relevant information, which is to use the program "strings." When you point it at a file, it will extract and list all consecutive printable characters longer than a given threshold. It produces a lot of gibberish, as there is a certain probability that random binary data produce a set of printable characters. However, even with tens of thousands of strings to scroll through, it is simple to spot the folder-path-looking string, which is exactly what happened in this case. We will return to the strings program in one of the other case examples.

With a probable username connected to the promoted proxy software, the question became what the chance was that this person also used this username services out on the Internet?

4.2.1.3 Accounts on Social Media

A search for the username using a popular search engine returned several results, including social media, forums, and online payment services. It became clear early on that this investigation may result in quite a bit of data and relationships, so we started adding the information we found as nodes and links in Maltego. We added each found service (forum, social media), username, display name, and email addresses as nodes and links between them where they were found together.

One display name appeared connected to the username on multiple services. The name was common enough that a search on Google for it returned a lot of results, but

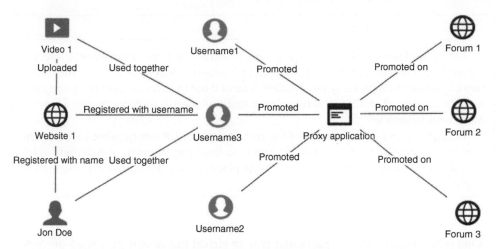

Figure 4.1 Illustration of investigation network for proxy application.

one of the top ones was from a known video sharing service. The result was from a user page for a user with the same display name as we found connected to the presumed username of our proxy seller, and with a matching username, only with a two-digit number appended to it. It seemed likely that this account belonged to our proxy seller (Figure 4.1).

The videos the user had uploaded were related to a completely unrelated subject. There was, however, one video that looked different. In this video, presumed addresses of the user, as well as other identifiable information, could be observed.

4.2.2 The Scammer

In this next example, a number of a suspect's computers and phones were seized. The suspicion was that the person had been making attempts at defrauding customers of an online bank by gaining illegitimate access. On the seized devices, chat logs where the person seemed to be discussing the matter with several individuals only referenced by alias were quickly identified. A number of virtual machines (actual operating systems running within the system, not the program type discussed in the previous chapter) were present on the computers, including a type of virtual machine designed to provide "anti-detection" to remain anonymous.

Example 4.2: The scammer

A number of a suspect's computers and phones were seized. The suspicion was that the person had been making attempts at defrauding customers of an online bank by gaining illegitimate access. On the seized devices, chat logs where the person seemed to be discussing the matter with several individuals only referenced by alias were quickly identified. A number of virtual machines (actual operating systems running within the system, not the program type discussed in the previous chapter) were present on the computers, including a type of virtual machine designed to provide "anti-detection" to remain anonymous while conducting your business.

(Continued)

Example 4.2: (Continued)

One of the features of this particular type of virtual machine is password-protected browsing profiles. These browsing profiles take the form of encrypted zip files. There were two big problems with the approach: One, it turned out the service used the same password to encrypt all profiles. And two, virtual machines persist memory to disk when the machine is suspended.

Through analysis and extraction of the correct password, it was possible to unlock the files. Unlocking these profiles gave us access to browsing histories and caches, which would help in understanding how the other pieces found during the investigation fit together.

One of the features of this particular type of virtual machine is password-protected browsing profiles. These browsing profiles take the form of encrypted zip files, and it functions somewhat like this: The user authenticates with a password to the virtual machine provider and is then provided the option to open one of the existing profiles or create a brand new one. Once a session has been selected, the provider will decrypt the zip file using some password known only to the provider and open the browser using the contained data (like browser history and sessions.) One benefit of these sessions is that it makes it easier to manage multiple personas online (email, social media, web shops) simultaneously.

There were two big problems with the approach: One, it turned out the service used the same password to encrypt all profiles. And two, virtual machines persist memory to disk when the machine is suspended.

As previously mentioned, there were multiple virtual machines present on the seized computers. As these machines can be put in hibernation, we' were able to open them with any running programs still open. In one of the machines, there was an instance of a password-protected browser profile open. What this means is that the password-protected zip file backing the profile at some point was decrypted by the provider. For a password-protected zip file to be opened, its password must be loaded into memory. What' is the chance that the password is still present somewhere inside the running memory of the virtual machine?

The password of a password-protected (encrypted) zip file takes the form of a hash. If we extract the hash of one of the encrypted profiles, we have an efficient way to check if any string in a list is the password to unlock the profile. Our list of strings, in this case, is the output of running the program "strings" that we mentioned earlier on the persisted memory file from the virtual machine with a profile already open. By luck, in less than a second, "john the ripper," the tool we used to check extracted strings against the hash, finished outputting a valid password to unlock one of the encrypted archives. Up until this point, it was not known that all encrypted profiles would share the same password, but the found string looked something like "serviceprovider123", and sure enough: That password opened all encrypted profiles we could find.

Unlocking these profiles gave us access to browsing histories and caches, which would help in understanding how the other pieces found during the investigation fit together.

4.2.3 The Disgruntled Employee

A company had started outsourcing IT services like server maintenance. They had already outsourced the hosting of the company's website, but at times, the servers were being unresponsive. The vendor could confirm that their servers during periods were receiving large amounts of requests and were unable to successfully complete them all. There is a growing sense that the servers are being victims of a distributed denial of service (DDoS) attack: Thousands of likely hacked computers are told by some operator to start sending as many requests as they can to a specified target. There is not much at the time the vendor can do to distinguish the bad requests from the good ones, and their infrastructure was not built to handle this amount of traffic.

Example 4.3: The disgruntled employee

A company had started outsourcing IT services like server maintenance. They had already outsourced the hosting of the company's website, but at times the servers were being unresponsive. The vendor could confirm that their servers during periods were receiving large amounts of requests and were unable to successfully complete them all. There is a growing sense that the servers are being victims of a DDoS attack.

Through analysis of network logs and tracing of the involved IP addresses, we could determine likely users of a given address. One of the IP addresses points back to the network of one of the vendor's customers.

During the investigation, request logs from the relevant servers were collected. Logs from a DDoS attack will contain a large amount of noisy data. The logs usually contain information like timestamp, the request's path on the webserver (like /some/path.html), the IP address (which may be the last proxy), the response code, the request size, response size, server, and the response time. Information may be available depending on the logging configuration. Most of the IP addresses in the logs after a DDoS attack will trace back to compromised (hacked) computers that are part of a "botnet" and will likely provide few clues as to who is behind the attack.

The fact that most of the requests went to a specific path on the server made it possible to filter out a lot of the noise coming from the attack. The remaining IP addresses can be collected and analyzed further. The idea is to look for clues that someone is scouting on the website, for example, looking for parts of the site that require more resources to complete than others, making them better targets for the attack. This could end up looking like an enumeration of pages on the website, where one IP is opening every available page on the site (this may also be a "spider" from search engines that' is crawling the Internet for information to index).

The geolocation of the IP addresses may help understand where the requests are coming from. Further, we may want to trace network information about each IP address, for example, using the "host <ip address>" Unix command. For private individuals, this will likely return the name of the Internet provider, but in the case of companies that have domain name system (DNS) records pointing to their servers, the name of the company may appear in the results. With the proper authority, we could also request additional subscription information about the IP addresses from the Internet service providers (ISPs), which we could use to determine likely users of a given address. One of the IP addresses points back to the network of one of the vendor's customers (with reference to the insider threat actor as introduced in Section 1.2.2).

4.3 Networking Essentials

To be confident in your investigations on the Internet, it is important to have a good grip of how data is transferred between computers and how core IPs operate. The reader is encouraged to look for primers on computer networks, but we give a brief introduction here.

Each computer connected to a network has a media access control (MAC) address, which is a static identifier of the network card that does not change with the currently connected network. Each machine is also assigned an IP address, which is commonly either set to a fixed value by the user or assigned by the network router using Dynamic Host Configuration Protocol (DHCP). This IP address lets computers send data packets to each other. As these addresses may change, it is impractical to use these addresses as references. Instead, if the user is meant to be accessed by other computers, it is common to pair the IP address with a domain name using the DNS.

When someone, or something, wants to communicate with a certain computer, the first step is to look up the IP address in a DNS nameserver. This server is identified with a static IP address, and there are a few of them, like 8.8.8.8 managed by Google. But this nameserver does not contain information about all domain names – IP pairs. Instead, it looks at the domain name in parts, separated by dots. Take, for example, the domain stud.ntnu.no. Google does not know which IP address this domain is associated with, but it does know that ".no" is managed by nameserver i.nic.no. It then asks that nameserver for the IP address, but i.nic.no does not know either. However, it does know that .ntnu.no is managed by the server's name ns.ntnu.no. Asking ns.ntnu.no finally reveals the IP address for computer(s) associated with stud.ntnu.no. You can perform this check by running "dig +trace stud.ntnu.no" from the command line.

Now we want to communicate with this computer. This usually happens using IP version 4 or 6. Version 4 addresses are 32-bit integers separated into four numbers, each between 0 and 255, e.g. 192.168.0.1. Version 6 addresses are 64-bit integers and are represented with hex characters. (The IPv6 representation of the previous IPv4 address is 0:0:0:0:0:ffff:c0a8:1.) The data we want to send to the other computer is wrapped in a packet consisting of a header and a body. Among other things, the header contains the destination address. The packet is sent from the origin computer to the closest Internet router. This router usually has a number of *legs*, which it can send data to and from.

When the router receives a packet, it checks its internal configuration to determine which leg to pass the packet on to. This usually means sending the packet to a router managed by your ISP. This process is repeated until the destination is reached or the packet's time to live (TTL) is reached. The TTL is the maximum number of routers the packet is allowed to go through. When the ISP routers determine which leg, or route, to take to the destination, there may be economics and politics involved. ISPs negotiate deals between themselves to share infrastructure. Often smaller ISP pay larger ISPs to have them transfer their customers' packets, and other times they may agree that they both benefit from using each other's infrastructure.

The protocol used to determine routes between ISPs is often the Border Gateway Protocol (BGP). The protocol is designed so that changes made by one router are automatically propagated to the others, which may lead to something called BGP hijacking. One example of this is when the Pakistani government in 2008 wanted to block access to YouTube for its citizens but ended up blocking YouTube for the entire Internet (BBC News, 2008). This is often referred to as *blackholing*.

The IP address of your personal computer is most likely not globally unique. Instead, it is unique within your current local network. All devices connected to this network share the same IP address out on the Internet. When the router receives a packet that is intended for one of its currently connected devices, it checks its network address translation (NAT) table to determine where it should be sent. This works by assigning a source *port* to packets when they are sent out on the Internet. When the other computer responds with a packet to that port, the router knows where it originated.

Now that the two computers can communicate, transfer protocols like Transmission Control Protocol (TCP) and User Datagram Protocol (UDP) are used to enable application protocols like Hypertext Transfer Protocol (HTTP), FTP, SMB, and IMAP to function.

4.4 Networks and Applications

The layers computers use to communicate with each other are available to us during our investigations. We can figure out which IP address a domain is associated with or determine if an IP address is associated with any domain names. Applications listen on ports to let other computers communicate with it, so we can do port scanning to possibly identify applications running on a computer or on a network.

Port scanning is often associated with cyber intrusions, and while usually not in itself illegal, there is an example in Norway from 1998 where the scanner had to pay for damages made on the receiving end of the scanning. The reason is that the target of the scan had reason to believe that it was being attacked and began costly measures to counter this attack. Interestingly, the scan was part of a broadcast on Norwegian public television to illustrate vulnerabilities in information systems (Norman-dommen, 1998).

4.4.1 Operational Security

Before you start interacting with the target infrastructure to trace information, it is important to be aware of what information you leave behind on those systems. Just like we use logging and analysis to protect our own infrastructure from attacks, your targets may use the same techniques to detect investigators. If this information can be traced back to you, e.g., if you are using your own computer from your home, it can become uncomfortable. So, beware and use things like proxies, anonymization networks, and dedicated hardware for this kind of activity. The awareness and techniques used to protect activity are commonly referred to as *OpSec* (operational security).

4.4.2 Open Sources

Most information accessible through the Internet may be referred to as open sources. It includes search engines, social media, public records, and many other types of information. Because of the vast amount of available information, it is important to identify, collect, and systemize it in a structured manner.

Open sources are commonly used as a source of intelligence to help understand the infrastructure and capability of targets, but they can also be used for tactical purposes. One example is to discover that two seemingly unrelated websites are managed by the same entity because they' are using the same tracking identifiers, e.g. *Google Analytics*. Another is to map out the social circle around a target by crawling social media platforms like Facebook, Twitter, and LinkedIn, as well as online forums. By automating the collection of such data, you can also perform analysis on the data, like finding the coinciding activity of things like posts, follows, and account creation. We will explore open sources in more detail later in the chapter.

4.4.3 Closed Sources

Creating fake personas or reusing cookies from acquired evidence are two ways to gain access to closed sources, such as protected groups in social media. Websites use cookies to determine the identity of a visitor. This cookie is usually a long string or large number that is infeasible to guess. However, by digging into browser data within acquired evidence, we may find and transfer these values to another computer and reuse the active session. Please note that there are legal and ethical considerations to be made before conducting these kinds of operations.

Law enforcement and other organizations may have entire departments dedicated to getting access to closed sources on the Internet. Some access requires social engineering or other types of concealed activity to become a member of groups on social media, forums, etc. In many groups where the intent is to stay hidden and anonymous, it is common for members to require new members to be vouched for by one or more existing members, as well as requiring members to be active by providing original content. It can therefore require significant effort to gain and maintain access to these networks. It can, depending on the situation, however, be paramount to successful investigations.

4.4.4 Networks

Command-line tools like whois, ping, nslookup, masscan, and nmap can help you understand a computer network: Which computers are running and what applications they are running can be relevant to understand if this is part of someone's intended infrastructure, or if it is a compromised host that is being used as a proxy. Services like Shodan[1] have already done the scanning for you, so you do not have to do it yourself. Note that computers may be configured to only accept connections on a given port from a given IP address or IP range. So even if your scan is negative, that does not necessarily mean that nothing is running on that computer or on that port.

---

1 https://www.shodan.io

A less obvious way of tracking infrastructure is to send specially crafted requests to servers in order to have them unintentionally reveal information about themselves. One example of this was the scanning of SSH key reuse of applications running on Tor that was used to determine which services were managed by the same person or group. Another example is from an investigator in a Norwegian newspaper, who discovered that he could make a misconfigured web server running on Tor reveal its actual IP address by uploading a picture in a specific fashion (Mützel, 2017).

To understand your own network during an investigation, you can use DHCP and DNS logs. DHCP logs contain information about which computer was assigned which IP address at which time and passive DNS logs contain information about which IP address a DNS lookup resulted in. These logs can help identify compromised computers within a network, for example, when malware is sending requests back to their command-and-control servers. They can also help establish timelines for when hardware or software was introduced, used, and last seen on the network.

There are online tools that let you estimate where in the world a computer with a given IP address is located, for example MaxMind[2]. However, note this location is not fixed, as IP ranges may be moved around or even hijacked by intentionally or accidentally misusing BGP.

4.4.5 Peer-to-Peer

Peer-to-peer networks are networks where the members are connecting directly to each other, and the role of a central server is the discovery of other members. Common examples of such networks are BitTorrent and other file-sharing networks, as well as the network used to operate distributed databases like the Bitcoin blockchain, which we will discuss later in the chapter.

Every network may function a little differently, but normally when a new peer wants to connect to the network, it will bootstrap the connection process by connecting to one or more DNS *seeds*. These seeds are typically bundled with peer software. The initial request will then return a set of IP addresses of other peers that are connected to the network and that the new peer can connect to. These peers are in turn connected to other peers, and together, they make up the peer-to-peer network.

Due to how this connection process works, members put a certain amount of trust in each other as they open for connections from other computers. In particular, if the network is small and each peer is connected to a high percentage of all nodes in the network, this trust can be misused to trace other peers by their IP addresses. However, like with network applications, these IPs can be masked using virtual private networks (VPNs), proxies, or onion routing. Still, when computers are connected using some application protocol, it opens up the possibility of sending crafted messages that cause the application to misbehave, either by returning information it should not or, in the worst case, enabling the sender to execute arbitrary commands on the recipient machine.

---

2 https://www.maxmind.com

4.4.6 Applications

There are a *lot* of different applications connected to the Internet. As they are built to interact across a network, we can send them messages of our own choosing. And application is designed to be communicated with in a certain way, for example, with a specific message protocol or application programming interface (API). So, if we send random bytes, it will most likely respond with some message saying it did not understand.

By using the application in its intended fashion, for example using its website, we can inspect the messages going back and forth between the client and the server. The most appropriate technique may vary depending on the type of application. For a website, we can quickly inspect these messages, or requests, by opening the browser console (right-click on the website, click "inspect," and click on a tab called something like "network"). For other types of applications running on our computer, we can inspect messages by listening to network traffic using something like Wireshark[3]. If the messages are encrypted, we may still infer what is being sent back and forth by attaching a debugger like IDA Pro[4] to the running application.

4.5 Open-Source Intelligence (OSINT)

Open-source intelligence, or open-source intelligence (OSINT) (see Definition 1.12), is a term used to describe the process of leveraging data on the Internet that is available without special access or authorization. Both businesses and criminal enterprises use it to better know the opposition and keep track of their activities. This knowledge can enable better decision-making, guessing passwords or imitating people more efficiently, or finding information that was not necessarily meant for the public. With the vast and ever-expanding amount of data on the Internet, OSINT poses both unique possibilities and challenges. While it is possible to take advantage of the extra data, it is also easy to get lost.

4.5.1 Methodology

There' is a saying that if you' are doing OSINT without a plan, you' are just surfing the web. Clicking random links may lead you to interesting bits of information, but without a clear sense of how and why you got there, it can be hard to use and share information properly. The process for open-source intelligence is similar to other intelligence processes. We start with a clear mission statement by an intelligent customer of the work's goal. In this case, the goal is often to answer some specific question, like "is A and B friends?," "When did A last go to C?," etc. Based on this plan, we identify potential sources that can provide information to reach that goal and determine how these sources are best accessed.

With the relevant data collected, information is processed to connect the dots between collected data and our goals. Then these connected dots are presented in an

---

3 https://www.wireshark.org
4 https://hex-rays.com/ida-pro

analysis that does its best to answer the original question. This analysis is then disseminated to the intelligence customer, which provides any feedback on the result to improve the process further. We cover processing and analysis elsewhere in the chapter, so in the following, we focus on how to identify relevant open sources and how to determine how to retrieve data from them.

4.5.2   Types of Open-Source Data

We have already briefly discussed network data as a source for Internet investigations. Moving up to the application layer, we find all sources available over the HTTP. This includes all websites and most APIs that we can access to request information. Some websites are more frequently used than others, like content search engines like Google and DuckDuckGo, but also services indexing network information such as Shodan. Waybackmachine is a service specializing in indexing past versions of a website, which can be helpful to detect if some information should be hidden from the public.

Geospatial information such as vessel trackers, flight trackers, and satellite images can also be invaluable to finding and keeping track of objects relevant to our investigations. There are also public registers from around the world and services like Open Corporates that pre-index and combine data from multiple sources. Public records can contain information from organization and ownership structures to public tenders and weather data.

Some of these services allow us to access the data through APIs, enabling us to gather data in an automated fashion, which again may increase the efficiency, precision, and quality of our data collection. Often these APIs are available by purchasing a license, which may or may not be relevant for you or your employer.

4.5.3   Techniques for Gathering Open-Source Data

The browser is our primary tool for accessing open sources, and printing websites to portable document format (PDF) is one method to ensure the integrity of the data we collect. However, there are other techniques we can use to focus our data collection better.

4.5.3.1   Search Engines

Most are comfortable using search engines like Google or DuckDuckGo to find information. However, search syntax can be used to either expand or narrow our search results depending on our needs. A more complete guide for Google can be found in Footnote[5], but a few tricks the author has found handy include:

- "ext:pdf" search only for results that come from resources with a particular file extension
- "site:ntnu.no" search only for results that are indexed on a particular website
- "-something" excludes search results that contain the word "something."

The available search syntaxes that can be used will evolve with the service provider. Be aware that techniques may exist that meet your current requirements.

---

5 https://support.google.com/websearch/answer/2466433

4.5.3.2 Browser Plugins and Bookmarklets

Browsers can be extended with additional functionality to support data collection during investigations. Such tools, like Hunchly, are often built to keep track of pages visited, with content, and keeping track of steps taken to get to a certain point. This is particularly useful when you, at some point in an investigation, suddenly realize that something you observed earlier is a lot more relevant to the investigation than first anticipated. Other tools like Maltego[6] and Spiderfoot[7,8] seek to aid and automate data collection from other sources. Another potentially helpful technique is called bookmarklets. The browsers treat bookmarklets like regular bookmarks, but when clicked, some JavaScript code is executed. This can typically be used to send the current Uniform Resource Locator, or URL, to someplace you can retrieve it later, possibly with some additional context.

A simple service that uses bookmarklets to collect and organize data in open sources is made by the author and is called BookmarkIntelligence[9]. The tools let you create bookmarks for different object types, like a person, organization, location, bitcoin address, or event. As open sources are being explored, information can be stored by selecting some relevant text and clicking the bookmark. This will create objects within the service for the current website and the selected text and create a link between them. This makes it possible to easily keep track of where information was found and visually explore and create relationships between objects. Figure 4.2) is a screenshot from the service.

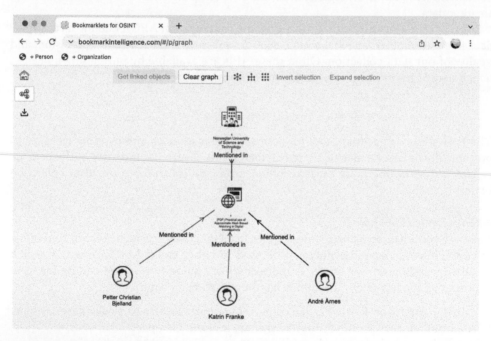

Figure 4.2 A screenshot of data gathered using BookmarkIntelligence (b18e.com).

---

6 https://www.maltego.com
7 https://www.spiderfoot.net
8 https://github.com/smicallef/spiderfoot
9 https://b18e.com

4.5.3.3 Scripting and APIs

Open sources can be accessed through other software than just the browser. Simple Python scripts send HTTP requests to servers, a common way to use APIs. It is also possible to use scripts to download websites as HyperText Markup Language (HTML). However, as modern websites often include many scripts being run within the websites, it can be tricky to make the downloaded content look like what it does in the browser.

A library called Selenium is one way to overcome this challenge. Selenium is a scriptable browser application that also can run in the background. What a website looks like in the browser may also depend on the cookies and other request parameters sent in the HTTP request. From the browser console (right-click somewhere on the website and click "inspect") under the "network" tab, you can see and copy requests sent to the server by right-clicking on the request of interest and clicking "Copy as cURL" (or similar).

Next, we will look more closely at Internet browsers, how they can be used and how they can be a source of information.

4.6 Internet Browsers

In addition to the information found on the Internet, we may also be in positions where we need to investigate user activity on the Internet from traces on acquired computers. In the following, we discuss the types of traces left on a computer from using an Internet browser and how to access them. We will also look at how Internet browsers can collect information and interact with servers beyond opening web pages.

4.6.1 HTTP, HTML, JavaScript, and Cache

Internet browsers are developed to be as user-friendly as possible. A result of this is that the browser will try to remember everywhere a user has been for them to go back later. When a user loads a website, the webserver typically first sends back a text file in a format called the HTML. The browser will interpret and present this file to the user, and as part of the rendering, see if other resources should be loaded from that web server or others. These types of resources are typically images, stylesheets, or JavaScript files. In modern web applications, the initial HTML file is usually tiny, with a loaded script being responsible for manipulating the initially rendered website into the actual application. In this process, additional resources may be downloaded. It is common for search engines to include the search terms within the URL, which means that search history also may be available as part of the user's browser history.

Browsers also try to minimize the amount of time spent downloading resources by storing them more permanently and retrieving them from local storage rather than the server next time. If the same large picture, for example, is downloaded on every page, it may be enough for the browser to download it once. These cache entries are typically referenced by the URL from where they were downloaded. An example of where this can be useful is when the picture of a person was downloaded from a URL that included that user's ID on the service. The interaction between client and server happens in the client requesting a specific image or JavaScript file. This can cause cache entries to provide information about how the user was interacting with the website, such as the user's search terms.

4.6.2 Uniform Resource Locators (URLs)

The previous section discussed how resources are cached by the browser and referenced by the resource's URL. A URL can consist of many different parts, but normally they consist of the following:

```
<protocol>://<domain>:<port>/<path>?<param1>=<value1>&amp;<param2>
=<value2>#/<hash>
```

The protocol is often typically "HTTP" or "HTTPS." The domain is used to identify the server's IP address, and the port is used to the application on the particular server. Note that when the protocol is HTTP, the port is assumed to be 80 unless specified, and if the protocol is HTTPS, the port is assumed to be 443. The URL path is used to locate the resource that is being requested, but so are the query parameters and their values. The URL's hash usually is not sent to the server but is used by the browser to maintain some state. Server applications are free to choose how they want the URLs after the port to look, so you may come across URLs where everything after the third forward slash is a seemingly random string.

4.6.3 Cookies and Local Storage

In addition to the URL, the browser also sends several *headers* together with the request. These header fields tell the server how the client wishes to receive the response and tell the server who the user is. The latter is usually done using *cookies*, a collection of key/value pairs stored by the browser within the current domain or path scope. When a user logs into a web application, its credentials are typically sent to the server for verification. If the server accepts the credentials, it will return a cookie value that the browser can send in future requests rather than sending the credentials for each request. In Chromium-based browsers (Google Chrome, Chromium, Edge), cookies for an application can be viewed by opening developer tools and then going to the "application" tab and selecting cookies.

In addition to caching resources and storing cookies, modern web browsers also include some application storage capabilities. In these stores, applications can store data for offline use or similar. Suppose it is a specific web service relevant to the investigation, and it seems like that service is actively using local storage in browsers to operate. In that case, it may be worthwhile looking into extracting and parsing this data to use for analysis.

4.6.4 Developer Tools

We have briefly mentioned the browser developer tools. As the name suggests, it is primarily a tool made to help developers debug and analyze their applications, but it can also be helpful for investigators. While developer tools are available in most browsers, we will assume a Chromium-based browser, which means browsers like Google Chrome, Chromium, and Edge. The easiest way to open developer tools is to right-click somewhere on the relevant web page and click "inspect." You can also open through the browser menu. Three tabs may be particularly suitable for investigators, which are "application," "network," and "console."

4.6.4.1 Application

The application tab shows the browser's data for that particular website. This includes stored cookies and local storage. Suppose you are allowed to do so, and you have extracted a user's cookies from forensic evidence for that particular service. In that case, you may be able to resurrect the user's session on that website by changing the cookies to the ones found in evidence. This could, for example, be done to access someone's email account or social media profile. There are obvious legal, ethical, and OpSec concerns to consider here, so make sure you are authorized before doing this.

4.6.4.2 Network

Under the network tab, you can observe requests going from the website to different servers. These requests will contain several downloads of packed javascript and stylesheets, icons, and images, but also stylesheets, icons, images, and application data. Clicking on one of the requests, you can observe the request sent to the server and the corresponding response data and headers. This information may provide unique insight into how an application is functioning. If you right-click on the request, you can copy the request data as a command that can run outside the browser.

4.6.4.3 Console

The browser console is commonly used to output logs and error messages from the application itself. Still, it can also be used to send commands to the application and fetch data from the server as if it was coming from the application. This can be useful to understand better how an application functions and how collected information may be interpreted.

4.6.5 Forensic Tools

Browser history, cache, cookies, and local storage are usually stored somewhere on the user's computer under one or more different user profiles. There are plenty of forensic tools that support extracting data from these files and profiles, for example Plaso[10], which is an open-source forensic timelining tool.

4.7 Cryptocurrencies

The history of cryptocurrency predates the now-famous white paper about Bitcoin published under the pseudonym Satoshi Nakamoto in 2009 (Nakamoto, 2009). Hashcash, an earlier version of the concept, was suggested already in 1997 (Back, 1997). It was, however, the introduction of Bitcoin that cryptocurrencies started gaining traction. There are several ways they can become relevant for investigations, including

---

10 https://github.com/log2timeline/plaso

employees misusing employer infrastructure to mine cryptocurrencies for-profit, people using cryptocurrency to pay for illicit goods, groups receiving payments from victims of ransomware to recover data, and money laundering.

In the following, we focus on Bitcoin, but the content should be applicable to the ever-expanding selection of available cryptocurrencies. The following is based on surveying existing literature and observations from engaging with the Bitcoin network. Given a public key, a bitcoin address, it' is possible to prove you possess the private key if you have it. When proved, you can tell the network you want to spend those funds associated with that address to another address. A set of transactions is combined into a "block" by one of the computers connected to the network or miners. When the miner has solved an equation based on the transactions, it adds the block to its ledger or blockchain and broadcasts the new block to their connected peers. It is trivial for these peers to verify the validity of this newly mined block, and if verified, they will drop any current equations they are working on and start over based on the new block (Atzei *et al.*, 2018). In cases where multiple nodes find a solution to a collection of transactions, the earliest one is usually selected. Transactions that went into the solved block that did not make it onto the global blockchain need to be attempted and verified again. The fastest miner gets a cut of the approved transactions called the fee.

4.7.1 Addresses and Transactions

A bitcoin transaction is a chunk of data added to the blockchain by the connected network. Within a chunk of data is several attributes required for performing the transactions. Attributes of interest include the input and output Bitcoin addresses, transaction date, and output amounts. The included addresses are public keys, and only inputs that are signed with the corresponding private keys are accepted. The output addresses are also public keys pre-generated by the recipient. Inputs to transactions are typically referenced by the transaction ID and output index from which the funds arrived.

When an input address is accepted in a transaction, the entire amount available on the input is emptied. Each input has corresponding output from where the funds came, except for mined bitcoins from the network itself (the "coin base"). An address may have received multiple transactions, and each of these "unspent outputs" is associated with the recipient address. When the recipient address makes a transaction, one or more of the unspent outputs will be used as inputs. This explains why you can see transactions with multiple inputs from the same address. The exchange is returned to the sender to a freshly generated change-address.

A single transaction might contain multiple outputs if the sender added multiple recipients to the same transaction.

4.7.2 Privacy

Before we look at possible methods and scenarios for grouping addresses by the assumed owner, it is essential to note that it can be very hard. Remember to always look for ways to disprove your hypothesis and combine your findings with data from other sources.

While the Bitcoin blockchain with all its transactions is transparent in that anyone can download the entire thing, it is also built to make it hard to link transactions to individuals. For starters, there is no personal information on the bitcoin blockchain.

With enough hardware and assuming a wallet owner used bitcoin locally, it is theoretically possible to link a transaction to an IP address by tracking who first publishes the transaction. However, as each node only is connected to a handful of other nodes, it is assumed to be very difficult. Not only because of the amount of computing power required but also because nodes can run connected through VPNs or onion routers.

A more feasible approach is detecting when funds are moved to an exchange like Coinbase or Binance to be converted to fiat currency. Bitcoin has a couple of privacy features that make this monitoring harder to succeed. Going back to how transactions are designed to work, we have a couple of different transaction scenarios we may encounter.

4.7.2.1 Single Input, Single Output
The entire amount was sent to exactly one recipient. We do not know if the sender and recipient are the same entity.

4.7.2.2 Multiple Input, Single Output
The sum of funds from many addresses is sent to one recipient. We do not know for sure, but it is likely that the same entity controls all sender addresses. If several inputs come from the same address, the address has received funds in multiple transactions. This is something that is usually avoided for privacy reasons, so the presence of this may provide some signal about the owner of the funds.

4.7.2.3 Single Input, Multiple Outputs
The funds of a single address are transferred to multiple addresses. If there are two recipient addresses, one likely scenario is that one entity sent bitcoins to a single other entity and that the second output address is a new change-address of the sender.

4.7.2.4 Multiple Input, Multiple Outputs
A transaction may have multiple inputs and outputs. While all addresses may belong to a single entity or two, it may be tough to determine. It is likely not possible to determine with certainty which input address sent funds to which output address. This type of transaction is often seen when users buy bitcoin and exchanges, where the exchange bundle purchases from multiple customers into a single transaction. A similar pattern can be observed when customers convert their bitcoins into fiat currency.

As transactions with both multiple inputs and outputs provide an extra level of privacy, it has been attempted to be introduced with techniques like "coinjoin," where multiple transactions are combined into one to break the linkage between input and output.

4.7.3 Heuristics

We can use certain assumptions or heuristics to group addresses. The purpose is to guess that the same entity controls two addresses so that we can get a simpler view of how funds have been moving between entities rather than addresses. There are two common heuristics: Common input ownership and change address (Meiklejohn *et al.*, 2013).

If a transaction has multiple inputs, three possible things may have happened: Multiple entities have gone together to create a typical transaction, a technique often referred to as "coin join," a single entity has sent a payment too large for a single of its addresses to handle, thus needing to send funds from two or more addresses, or a combination of the two. That a single entity sends funds from multiple addresses is called common input ownership. The number of outputs may give some extra signal about which of the two you are looking at. More outputs than inputs can be an indication of common input ownership.

If the transaction amount is less than the balance of the input, the "change" is sent back to the sender on a (usually) newly created address, in a scenario where a transaction has a single input and two outputs. It is possible that one of them is the change address and is controlled by the sender. One way to determine whether one of the outputs may be a change address is to see if only one of them has been observed before. This can probably be generalized to more outputs, where if only one of the addresses has not been seen before, it is a change address candidate. There are other techniques, for example, looking for round numbers in the other outputs and looking for exclusively matching script types.

Clustering bitcoin addresses into groups that are potentially controlled by the same entity is an open research area and where significant progress likely would cause changes to the cryptocurrency protocol and applications. One example is that it was not common to create a new change address for each transaction but rather send it back to the sender address. This made addresses a lot more permanent than today and therefore easier to keep track of for investigators. It is trivial to create a new change address for each transaction, and now it is common practice to create new addresses for every transaction to help protect the sender's privacy.

4.7.4 Exploring Transactions

There are plenty of blockchain explorers available online. You typically search for a transaction or an address on these and see relevant transactions and balances. A few apply any kind of heuristics to the data, making it hard to get a good overview. Commercial tools exist that try to provide higher-level overviews by grouping addresses and having dedicated personnel analyzing and labeling addresses. The latter is helpful to detect connections to known entities like exchanges, known cybercrime groups, etc.

4.8 Preparation for Analysis

Like information collected from any other sources, data we collect related to the Internet must be processed and organized for us to be able to make good use of it. There may also be additional information we want to extract, such as metadata from encoded data, query parameters from URLs, and entities from text. We describe some techniques that may help build a toolbox for such data preparation. The methods may help organize collected data into piles that can be analyzed separately to keep an overview.

4.8.1 Entity Extraction

Sometimes we are not sure what we are looking for within the extracted text, and there may be too much text to read through it all. Then it can be valuable to do another round of processing on this, where we extract *things,* entities from the text. Here we categorize these entities into *regular* and *named* entities. Regular entities can be extracted using *regular expressions,* often referred to as *regexes.* Regular expressions are patterns that define a finite state machine and are commonly used for string search. If a given input causes the state machine to reach an end state, we have found a match. These patterns can get quite hairy quickly, and there is a reason for the saying "I used to have a problem and decided to solve it with regex. Now I have two problems." A simple example, where we want to find numbers that are at least six digits long, may look like this in Python:

```python
import re
pattern = "\d{6,}"
string = "abc 1234567 def"
for match in re.findall(pattern, string):
    print(match)
```

In the above example, \d{6,} is the regular expression. \d means digit, and {6,} means at least six characters. Instead of \d, we could also have used [0–9]. Typical expressions can be used to find, including IP addresses, phone numbers, email addresses, bank account numbers, cryptocurrency addresses, and known date formats.

Not all entities are regular, meaning that no pattern clearly defines their structure. Typical such things include names of people and organizations and locations and street addresses. Suppose we are familiar with how these entities occur, which may be the case in large or frequent reports. In that case, we can still write a sufficiently accurate regular expression, but this is usually not the case for unknown text. Instead, we may want to rely on a statistical technique called *named entity recognition* (NER). Tools that implement this technique evaluate the probability that a given substring in a text is an entity of a given type. While the result of these tools may be impressive, they have the obvious limitation that we do not know what it missed, as well as that they are usually trained for a specific language, most commonly English. Stanford maintains one popular tool for doing NER 19 and is accessible through the Natural Language Toolkit (NLTK)[11], which is written in Python. Other available libraries for doing NER are HuggingFace[12] and SpaCy[13]. Both libraries have published trained models for languages other than English.

4.8.2 Machine Translation and Transliteration

If you have ever tried using a machine translation service like Google Translate to translate a text from language A to B and back to A again, you know how unreliable machine translation can be. Agencies and organizations have dedicated language translation personnel that can provide high-quality translations of text when needed. However, human resources are scarce, and machine translation can be helpful to triage which information should be sent to human translators first. When working with non-Latin alphabets, we may also extract some signals by *transliterating* text to Latin. Transliteration

---

is the process of mapping one alphabet to another and can be helpful to detect references to proper nouns like names of people, organizations, and acronyms.

4.8.3 Metadata Extraction

The information generated for transmission by computers can be encoded into some representation that removes formatting. One example of this is Pretty Good Privacy (PGP) keys. A typical PGP public key looks something like this (Figure 4.3):

```
-----BEGIN PGP PUBLIC KEY BLOCK-----
Version: GnuPG v2

mQENBFYrQjsBCADuiShsojhBT7NUA5F3prD016E6MhJ82S0HmbNgyptNNPiSW0al
rHcFF6/cEWpkpnn2uI7rFlW17vnWRqRva3D1CbxmWu+MawRIxokt01UIfX3w7tEe
... (20 lines)...
3A91zHHahSF1WpUkIb3MLC8lPMZBwjl1z2rlHEcU1wkax6wyCcLkjtM2EJiCGzGp
wJqjCDL4iTml/Isbxh8uUfABSUS8+NTIjQUduxyHaCbDwn+U
=h9gd
-----END PGP PUBLIC KEY BLOCK-----
```

Figure 4.3 A typical PGP public key.

And is commonly sent out to those the owner of the private wishes to receive encrypted messages from. It may look like a random string but parsing this chunk of *Base64* encoded text reveals metadata entered by the creator, such as the provided name and email address. Parsing the above with a Python script[14] produces the following output (Figure 4.4):

```
[
  {
        "signed_by_key_id": "4FA93D42",
        "signature_type": "subkey binding",
        "user_id": "Petter Bjelland <post@pcbje.com>",
        "created": "2015-10-24T08:32:59Z",
        "expires": "never",
        "signed": "2015-10-24T08:32:59Z",
        "encryption_algorithm": "rsa",
        "version": 4,
        "signed_by": "53A921664FA93D42",
        "fingerprint": "5C73CE14B408A8EE599511EB53A921664FA93D42",
        "key_id": "4FA93D42",
        "key_length": 2048
  }
]
```

Figure 4.4 Decoding the Base64 encoded metadata of a PGP key.

---

14 https://github.com/pcbje/parse_pgp_keys

There is, of course, nothing preventing the creator from entering bogus information. However, it may still be a source of intelligence, like email or nickname reuse may provide clues about links between different online accounts if you can combine this metadata with information from other sources, like bank customer information and transcripts.

4.8.4   Visualization and Analysis

There is no silver bullet to figuring things out from data during investigations. These "things" usually relate to the *5WH,* the five W's, and the H (see Section 1.1). These are *who, where, what, when, why,* and *how,* and there are tools and methods available to work with data that highlight information that may answer them. Most prominent are timeline, link analysis, and geospatial analysis, which are techniques you will become familiar with when performing investigation analysis. When working with these tools to answer the relevant questions, it is essential to have a structure to how the analysis is performed. It is easy for the analysis to become a "poking around" exercise and hoping that something will pop up.

4.8.4.1   Link Diagram Analysis (Graphs)

In most investigation series on television, there will at some point be several pictures of people pinned to a corkboard, with threads connecting them to communicate which individuals, groups, and events are connected. These graphs here referred to as *link diagrams,* help generate, maintain, and share complex knowledge about relevant targets. Each *node* (pin) in the diagram is an entity of pretty much any type: Hardware, IP addresses, mobile phones, people, companies, locations, etc. A *link* between two nodes shows that those two somehow are connected.

These diagrams can become huge, so we need to rely on techniques to make sense of them efficiently. There are many such techniques. Here we will discuss some of the *visual* and *statistical* ones.

4.8.4.1.1   Visual Techniques

Each node is presented with a particular style. In tools like Maltego[15] and Analyst's Notebook (ANB)[16], each node usually has an icon that is a figurative representation of the node type, as illustrated in Figure 4.5). If available, the figure can be replaced with a picture of the given entity. Beneath the figure or illustration is a *label* that tells the identity of that node, e.g. name or phone number. Other graph tools, like Gephi[17], nodes are by default a shape where the shape color can specify the node type. Each node can be further decorated with styles and sizes to further specify its properties.

Links between nodes can be styled to explain the relationship between the two entities. Links can be *directed* so that even if A is connected to B, B is not connected to A, and it can be styled with thickness, dashing, and arrow types to further specify properties. It is also common to assign an *edge label* to links that describe the kind of content of the relationship between the two nodes.

---

15 https://www.paterva.com/web7
16 https://i2group.com
17 https://gephi.org

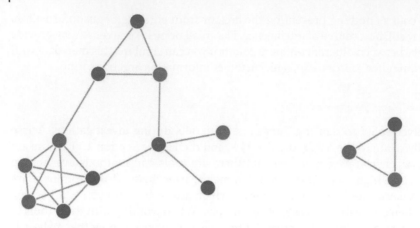

Figure 4.5 A link diagram (https://gransk.com/ggraph.html).

As diagrams become large, manually placing them becomes harder and harder. To avoid doing this manually, we can use one of the many layout algorithms that exist. Some common layouts include *hierarchical, circular,* and *organic.* The last one is also called *force-directed* layout, where each node has a specific simulated gravitational push on other nodes, while links create a gravitational pull between the linked nodes. The result is an organized layout that highlights *clusters,* which are groups of nodes that are more tightly connected than the others.

4.8.4.1.2 *Statistical Techniques*

Developed to understand better how people interact with each other, *social network analysis* (SNA) is a collection of measurements that predates *online* social networks by decades that are useful to determine essential nodes in a network. The importance of a node is often referred to as its *centrality,* and common centrality measures include *degree, closeness,* and *betweenness.* Degree centrality is computed by the number of links connected to a node. We can count the number of incoming, outgoing, or all links. The most central node is thus the one with the greatest number of links. Closeness centrality is computed by measuring how far the node is from all the others. The closest node in a connected graph is the one where the minimum number of *hops* (the number of links to traverse), to all the other nodes. The betweenness of a node *C* is computed by measuring how many shortest paths between other nodes go through C. The results of these centrality measurements can be fed back into the visual analysis by assigning node size and color based on their centrality.

4.8.4.2 Timeline Analysis

Organizing events chronologically is a valuable way to make sense of complex events. Events can simply be listed and filtered in a spreadsheet like Microsoft Excel. Tailored tools like Plaso[18] and Timesketch[19] have been developed to analyze digital evidence. Events may also be aggregated or grouped to view many events at once. Timelines can also contain relational components, where events are visualized as connections between entities, like the one illustrated in Figure 4.6.

---

18 https://github.com/log2timeline/plaso
19 https://timesketch.org

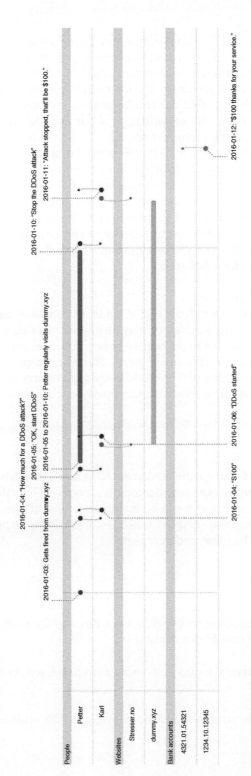

Figure 4.6 Relational timeline.

4.8.4.3 Pattern-of-Life Analysis

Though it may be hard to say for sure, we may sometimes have a suspicion that the same entity controls multiple accounts on or across some social networks. Suppose the activity is generated by a human and not automatically published by scripts. In that case, we may get clues to refute our hypotheses by looking at the accounts' pattern of life. The term pattern of life analysis is used to describe analysis techniques that seek to understand how an entity or collection of entities go about their day and is therefore often linked to categorizing temporal information.

In the case of social media, this may include what day of the week and what hour of the day posts are being published. Using the same time zone across is essential to spot differences in activity between someone sitting in Gjøvik, Norway, and someone sitting in Melbourne, Australia. If we put the hour of the day along the x-axis in a chart and the day of the week along the y-axis, we can create a heatmap comparing activity. If there is a clear and significantly different pattern between two entities, that may refute our hypothesis that they are the same. It' is important to note that this information can be deliberately generated to look different.

4.8.4.4 Geospatial Analysis

Many of the artifacts we extract from digital evidence contain geospatial information that can be plotted on a map. Each event may be plotted in a single shape, they can be grouped to highlight areas with a relatively large occurrence of events, or you can use a heatmap to highlight regions with many events. These heatmaps may be discrete (e.g., one color for each country, city, etc.) or a fluid range (e.g., colors). If the events also contain timestamps, it is possible to visualize movements or changes over time.

4.9 Summary

In this chapter, we have discussed the fundamentals of how the Internet is glued together and what data is available to us as investigators. We discussed how this information can be traced and acquired and how it can be interpreted. Several common analysis techniques related to link diagrams (graphs), timelines, and maps were presented to discuss how these analyses can be presented in reports.

4.10 Exercises

1   You geolocate an IP address on two different dates, and the results are widely different. What may have caused this change?

2   What are some potential issues with using automated scripts to collect evidence from the Internet?

3   What are some of the strengths and shortcomings of link analysis?

4 What are some of the strengths and shortcomings of timeline analysis?

5 How can a peer hide its IP address in a peer-to-peer network?

6 How can you determine if a peer is likely hiding its IP address?

7 Why can it be hard to keep track of Bitcoin transactions sent and received by a person, even if we know the person controls a specific Bitcoin address?

5

Anonymity and Forensics

Lasse Øverlier[1,2,] *

[1] *Department of Information Security and Communication Technology, Faculty of Information Technology and Electrical Engineering, Norwegian University of Science and Technology (NTNU), Trondheim, Norway*
[2] *Norwegian Defence Research Establishment (FFI), Kjeller, Norway*

From ancient times, we have built trust through the concept of meeting up in person to verify the authenticity of a message or action. At some point, we started to accept tokens, seals, monograms, and signatures, e.g., for verifying the authenticity of a sent message and for making binding contracts. This has led to the abuse of the same tokens, for-profit and other kinds of gain, and this has become a huge area of research within "analog" and digital forensics.

But being able to verify the authenticity and the origin of a message is not always the goal of a message. Some messages are constructed to be *anonymous*. Like hanging an unsigned political poster in a public place, there are online communities where political views and other types of information can be exchanged where the origin of the message is scrubbed and made anonymous. Anonymity has proven to be both a necessity and a problem for online communication in our current society. One easy-to-understand example is that the world has the need for information to be shared from the wrongdoings of oppressive regimes, but the oppressive regimes should not be able to identify the originator for obvious reasons. The other type of example can be seen every day with anonymous or fake accounts posting messages on the Internet containing threatening, false or illegal content.

We will, in this chapter, introduce the terms related to anonymity. We will discuss the already mentioned duality of anonymity, go deeper into how the technology is used to achieve anonymity, and give an overview of the limitations and strengths of the most used principles for anonymity. The entire chapter will address limitations and possibilities for doing forensics and investigations on these systems, and at the end, we will go through some of the possibilities for doing forensics involving anonymity systems.

---

* Lasse Øverlier is an Associate Professor at the Norwegian University of Science and Technology (NTNU).

Cyber Investigations: A Research Based Introduction for Advanced Studies, First Edition. Edited by André Årnes.
© 2023 John Wiley & Sons Ltd. Published 2023 by John Wiley & Sons Ltd.

5.1   Introduction

Most of the time spent through online communication, we, as regular Internet users, do not think about anonymity at all. But lately, more people are concerned with their privacy and the distribution of private information, which has become a major problem online. It is easy to relate to examples of information users expect to be private, but Internet users are often not able to see how to protect this information while using today's online services. Therefore, there has been a push for better and easier to use anonymity services, and this community has grown with the demand for online privacy. But privacy and anonymity are a double-edged sword and can also be abused by criminals and thus used to achieve protection from being traced online, a task which earlier was difficult or expensive. Today cheaper or even free tools will provide anonymity to a larger group of potential criminals and have too often been abused for this purpose. We will start the exploration of anonymity online by going through the most important principles of anonymity in online systems and how they work.

5.1.1   Anonymity

Anonymity means that others are unable to tell who did what. The term anonymity comes from the Greek "anonymia," which means "without a name." More formally, the definition can be related to a definition by Pfitzmann and Hansen (2007):

Definition 5.1: Anonymity and anonymity set

Anonymity is the state of being not identifiable within a set of all possible acting subjects, called the *anonymity set*.

This means that anonymity is simply a probability of being identifiable and that this probability is never zero. The probability of being identified is very low only if the anonymity set is very large. For example, "All we know is that a (single) human being did this." In this case, the anonymity set is around seven billion, and the risk of being identified is relatively small unless we have other forensic evidence that can reduce the anonymity set.

An *identity* is, in this chapter, used as a term for one (1) person participating in communication with another identity or towards a service of some kind, e.g., social media networks, web site, etc.

Definition 5.2: Identity

Identity is used as a term for one (1) person participating in communication with another identity or towards a service of some kind.

Often people think of cryptography as the same as anonymity, but as we can derive from the definition above, this is not the case. Cryptography, as most people relate to it, is used for hiding the content of the communication. But even if the content is hidden from external observers, the communicating parties are identified as the endpoints of the communication, so there is not necessarily anonymity involved.

This leads us to the term *traffic flow confidentiality*, TFC, which can be seen as protection of the flow of the content, not protection of the content itself (even if this normally must be present as well to achieve TFC). TFC is more commonly used in military settings where the protection of communication patterns is more important. For example, is it possible to derive the location of the sensors, weapon systems, and command headquarter just from analyzing traffic flow? This is, in essence, the same problem as anonymity systems try to hide.

Definition 5.3: Traffic flow confidentiality (TFC)

The hiding of the origin, path, and endpoint of messages/data.

When we look at message communication anonymity, we often separate *sender anonymity*, the anonymity of the origin of the message, from *receiver anonymity*, protecting the destination of the message, and *relationship anonymity*, protection from linking a sender with a receiver. Most anonymous communication systems focus on sender anonymity, as this is where they protect the person publishing the sensitive information in the first place. There is also the term *strong anonymity*, where two separate actions are prevented from being linked to the same identity.

Forward anonymity is when the compromise of long-term protection, like an encryption key, does not expose the anonymity in earlier communication. This is the same concept as with (perfect) forward secrecy (PFS).

Definition 5.4: Forward anonymity

Forward anonymity is when the compromise of long-term protection, like an encryption key, does not expose the anonymity in earlier communication.

Another term often confused with anonymity is privacy. *Privacy* is the ability to be in control of what personal information can be distributed to whom, and for online information, this now covers everything that can be used in a process to identify an individual: name, relationships, friends, phone numbers, surfing habits, images, bank account information, political views, etc. As we all understand, it has become an impossible task to have complete privacy – everyone must exist in some registers. You may choose to exist in some registries, but many companies and organizations build databases of your private information without informing you, and then you have no control of the information distribution. So, we see that privacy is the challenge of getting in control of the access and distribution of personal and private information.

Definition 5.5: Privacy

Privacy is the ability to be in control of what personal information can be distributed to whom.

It is probably more correct to look at privacy as a scale like we have some degree of privacy. There are many elements that will contribute to our degree of privacy, like our restrictiveness/paranoia, the trustworthiness of others to protect the information you expect them to protect, the integrity of the services you use, and the privacy laws and regulation of the country you live in or have relations with.

Privacy-enhancing technologies (PET) is a collective term used in privacy research that also covers research on anonymization technologies, including security related to both privacy and anonymity. As expected, this is an area of research where there is a lot of activity today, and there are many articles and books describing the early technologies of PET (Goldberg *et al.*, 1997) and later (Goldberg, 2002), and some recent publications on taxonomy for PET (Kaaniche *et al.*, 2020). We will cover some of these technologies related to anonymity protections later in this chapter.

We will not look at the challenge of measuring the level of privacy or the challenges of protecting the dissemination of private information in this chapter. But we will focus on online anonymity systems, which can be used as building blocks for achieving a higher degree of privacy in specific circumstances.

It is also a common misunderstanding to believe that the use of *pseudonyms* is the same as being anonymous. In some situations, a pseudonym can give the effect of being anonymous, but it is more correct to look at pseudonyms as a role given/taken for a period, in which period it may often not be possible to connect the pseudonym to the user's real identity. Some social media networks and online network forums give this possibility to their users, but the services themselves are (most often[1]) able to connect the pseudonym to the real identity, or at least to another pseudonym.

Anonymous communication is whenever we have a certain level of sender or receiver anonymity. What this level must be to achieve the wanted degree of anonymity (see next section) is up to the person seeking anonymity to decide. If the user is happy with being one of only two people, this may be enough in some cases, like two siblings blaming each other for something happening in the real world. But if this was a dissident trying to avoid being identified by an oppressive regime, this would only endanger both or maybe the entire family, and we understand that the anonymity set in this case must be larger to protect him or her. Maybe the anonymity set is not large enough, even when being one individual from a specific village or one from an identified region? Therefore, it is so important for anonymity systems to include as huge a variation as possible into the anonymity set. Not only because of location but also in all the other aspects of information that can be used for identification.

5.1.2 Degree of Anonymity

Is it possible to quantify and/or measure the level of anonymity that can be achieved in an anonymity system?

In a perfect system with an anonymity set of n participants, some defined degree of anonymity can be said to be like one minus the probability of exposure, $P = 1 - 1/n$. This means that when the anonymity set is growing, the highest degree of anonymity approaches one.

---

1 Many services must keep a record of customers either by legal requirements or to work properly.

There are many textual scales for anonymity. One "degree of anonymity" was presented by Reiter and Rubin (1998) and went from absolute privacy through degrees of innocence and suspicion to provably exposed.

It is also possible to think of anonymity as the system's entropy after an attack divided by the entropy before the attack (Diaz & Preneel, 2004). How to measure the entropy of anonymity systems is still an ongoing research area.

If you have a mixer network (see Section 5.1.2.4) of m mixers in a cascade, Berthold *et al.* defined the mixer network as secure if at least one of the mixer nodes could be trusted (Berthold *et al.*, 2000b). This still limits the anonymity set to the set of users of that specific anonymizing system.

The Nymity Slider, introduced by Goldberg (2000), gave a scale of anonymity ranging from *verinymity*, meaning the identity can be proved to *unlinkable anonymity* where the identity cannot be recovered. This was a concept that can easily be related to by users and is also used as a "risk slider" by the Tor project (Dingledine *et al.*, 2004) in their Tor Browser, now upgraded to a "Security level." The more potential vulnerabilities the browser exposes, for example, by allowing scripts, fonts, and playing media files, the higher risk of losing anonymity (Figure 5.1).

Through these security levels, the users of the Tor anonymity network can set their own preferred level of security in the Tor Browser. Even if this is not directly related to an easily calculated degree of anonymity, all potential security breaches may compromise anonymity as well. Therefore, the terms can be said to be correlated.

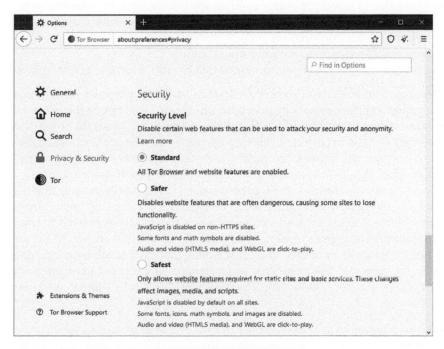

Figure 5.1 Tor Browser's security level (https://torproject.org).

In our simplified perfect Tor network[2], the compromise of anonymity is still easy (Øverlier & Syverson, 2006b) when an attacker controls both the entry and the exit node of the network. So, if an attacker controls c of the n servers in the Tor network, the probability of compromise is $(c/n)^2$, and therefore the probability of being secure is $P = 1 - (c/n)^2$ if all nodes are selected with equal probability. We will later see how the current Tor network differs from this, and how this could be exploited in locating Tor users and hidden services.

In the rest of the chapter, we will use the term "to be anonymous" when we describe a situation where the user has some degree of anonymity. We do not claim that the user is completely anonymous unless explicitly stated.

5.2   Anonymous Communication Technologies

There are many terms describing anonymity in online communications because we tend to describe anonymity in different ways related to where it is used or needed.

Anonymity is a hard term to relate to. In media, the term anonymity is most often used to describe an unknown perpetrator with malicious or criminal intent or activities. We will get back to some of the criminal scenarios of anonymity later in the chapter, but there are also many legal and necessary uses of anonymous communication[3]:

— Source protection for journalists
— Anonymous tips to the police
— Political dissidents publishing information
— Police investigation
— Personal protection of privacy in online communication, e.g., general interests, political views, health-related or economic questions, etc.
— Competitive analysis for corporations
— Procurement or research protection for corporations
— Business and government travelers communicating back to their base of operations

There also exists a lot of abuse of anonymity systems for illegal purposes. If you build a system to protect legitimate users by providing anonymity, it is impossible to avoid having the same system being abused for illegitimate purposes. This would be equivalent to saying that we need to have cars, but they cannot be used by criminals for criminal activity (or using mobile phones, computers, and so on). But this still does not mean that it is OK to abuse anonymity systems to perform criminal activities.

As with phones and cars, the laws and regulations need to reduce the damage that the criminals' activity can perform and at the same time protect society's legitimate needs for anonymity. This already exists with using cars where the police are allowed to trace car activity (with various degrees of depth and methods in different countries) on the background of, for example, by model, size, color, license plate, toll booth crossings, and traffic monitoring cameras. The same type of challenge will also exist for a forensics expert when the police are conducting investigations, have confiscated computer

---

2 "Simplified perfect Tor network": Excludes all other forms of compromise except endpoint traffic matching/correlation.
3 More examples are given at https://www.torproject.org

equipment, etc. Therefore, it is necessary for an investigator to understand in-depth how the different anonymous communication systems work.

In the rest of this section, we will look at some of the different types of anonymizing techniques, starting back at the origin with high-latency mixers and up to the more common anonymizing systems in use today, which requires anonymity and relatively low latency.

With *high latency*, we will address interaction which involves the sending of messages, emails, and publication of other types of information that does not require immediate feedback or response. With *low latency*, we describe systems where there is a requirement of near-immediate two-way communication, like normal Internet surfing and use, voice communication, live chat, and many social media live services. We will start by describing the high-latency systems as these were the first to be addressed in research.

5.2.1 High-Latency Anonymity

The start of high-latency anonymity was David Chaum's paper (Chaum, 1981) describing email mixes. An email mixer is a computer that accepts a large number of emails into the system, introduces new attributes to the emails, and sends them out again. These new attributes can be advanced like new cryptographic layers and new appearances, or simple new attributes like new random ordering of output email and or new identifying email headers. The latter was the case for Johan Helsingius' *anon.penet.fi* [4], maybe the most famous remailer of the early Internet (Helmers, 1997).

> ### Example 5.1: anon.penet.fi
>
> The service *anon.penet.fi* stripped off all identifying headers of the incoming emails and replaced the "From" field with an email alias named *anNNNNN@anon.penet.fi*, and forwarded the email to the recipient.
>
> The mapping between the real sender and the alias was stored in a database at the server, which made this server a single point of failure, even if it was essential that the server never kept logs of incoming and outgoing emails. This was called a *type 0 remailer* or *nym remailer* (for "pseudonym remailer"). This single point can be monitored, and it would be quite easy to map incoming names and outgoing aliases for the early (mostly unencrypted) emails, especially since the service did not change the content of the messages.
>
> In addition, it is easy to put pressure on the service provider if there are incidents that may break the (usually undefined) laws of these early pioneer services. The remailer service *anon.penet.fi* was closed[5] after legal pressure to get the real identities of users of the service.

---

4 https://web.archive.org/web/20060827210638/http:/waste.informatik.hu-berlin.de/Grassmuck/Texts/remailer.html
5 https://web.archive.org/web/20050921022510/http://www.eff.org/Privacy/Anonymity/960923_penet_injunction.announce

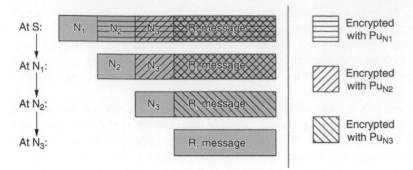

Figure 5.2 Message appearance at anonymizing network nodes in a sequence of mixes w. public-key cryptography (Øverlier, 2007).

Type I remailers, originating from the Cypherpunk community and therefore called *cypherpunk remailers*, were more complex and used a network of mixer nodes to have better availability and anonymity. The messages would add the recipient address into a specially designed email, encrypt this message with Pretty Good Privacy (PGP), and send it to the remailers. The remailers decrypted the messages, found the recipient's address, and forwarded the email to the recipient without keeping any logs of the transaction.

Type I remailers could also be chained by using public-key cryptography, as shown in Figure 5.2. Here the message and the identity of the recipient R are encrypted with the public key of Node N_3, then it adds the identity of N_3 to the message, encrypts the whole message with the public key of N_2, and repeats this with Node N_1. So, when the message is to be sent from the sender S, the observer can only see encrypted content going to node N_1. From N_1, the content is still encrypted but does not look the same due to the effect of layered encryption. The same goes for after N_2 and when receiving at N_3. If both users are using PGP on the inner content, it will be hard to map the sender message at S to the message going out from N_3 to the recipient R. This principle is the same as the concept of *onion routing*, which we will cover later in this chapter.

Type II remailers, also called *Mixmaster remailers*, use the same chained principle described above, and they are implemented in the systems Mixmaster (Möller *et al.*, 2004) and Babel (Gülcü & Tsudik, 1996). The major enhancements are improvement in resistance against message length matching and replay attacks, but there exist attacks against type II remailers, like the *n − 1 attack* (Danezis & Sassaman, 2003) and the *trickle attack* (Serjantov *et al.*, 2002).

Type III remailers, also called *Mixminion remailers*, add forward anonymity and long-term pseudonyms to improve usability and address the known attacks on type II remailers. The remailer network Mixminion (Danezis *et al.*, 2003) uses both key rotation, dummy traffic, and other improvements to increase anonymity.

5.2.2 Low-Latency Anonymity

When there is a need for anonymity in combination with low-latency services, like web browsing and VoIP calls, the systems described for message rewriting and remailers will not be sufficient. With low latency systems, many of the basic ideas and building blocks (encryption principles) are the same, only changed and optimized to fit into (mostly) two-way communication with higher interaction rates.

When in need of low-latency two-way anonymous communication, it is easy to understand that the threat model will be quite different. Any vulnerabilities in such a system will have a significant impact on the degree of anonymity since the high interaction will give an attacker more and often continuous information about the senders and receivers of communication traffic. In the case of remailers, this could be as little as one short, encrypted message and no connection to other messages through the system. With faster communication, like anonymous web browsing, VoIP, or even anonymous streaming, the anonymous system must try to hide the traffic flow information of thousands of network packets online traveling to/from endpoints wanting to remain anonymous. We will look at some of the most used principles.

5.2.3 Anonymous Proxy

If you route all your traffic through another computer, a proxy, it will appear to the other side of the communication channel that the traffic originates at the proxy's IP address. This is similar to most network home routers using network address translation (NAT), mapping all users in your house to one external IP address. The principle is like nym remailers, remembering your real address on one side of the communication but representing you towards the service you are accessing.

Most of these anonymizing proxies allow their users to connect to it through encrypted channels making the incoming and outgoing traffic different and appear unrelated (Figure 5.3).

So, if the proxy is (1) outside of your home address, (2) used by thousands of users, and (3) encrypted traffic is used for connecting to the proxy – this seems to be a good solution. And there are hundreds of such anonymizing proxy services available, with quite low costs for subscribers (ranging from free up to maybe hundreds of dollars per year).

But there are several challenges with this design. The two most important are:

First, there is the same vulnerability as with *anon.penet.fi*. The anonymizing service is a *single point of failure, a single point of compromise,* and *a single point of attack*. The opera-

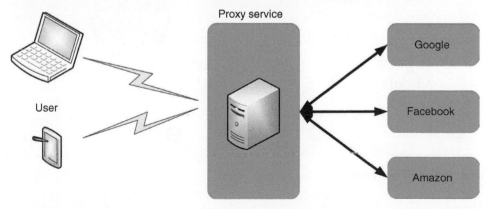

Figure 5.3 Anonymizing proxy. Low latency anonymity optimized to fit two-way communication with higher interaction rates.

tor of the service can (1) map each user to the accessed service, (2) may keep logs, may change, delete, and inject data into the communication channels unless end-to-end protected. Thus, the users will just have to put their trust in the proxy service provider in addition to the implementation of the software and hardware used in the service.

Second, it is relatively easy for an attacker with access to the network in and out of the anonymizing service to correlate traffic and map which encrypted connection is forwarded to what (potentially encrypted) server and vice versa. Most anonymizing services do not inject dummy traffic. Therefore, it is easy to look at packet sizes and data sent or received at the proxy over a period of time and compare the traffic patterns to connect the complete communication channels.

5.2.4 Cascading Proxies

And similarly to remailers, some people build anonymizing proxies in cascades, as shown in Figure 5.4, thereby enhancing the randomness of the system. JonDo[6], which has grown out of the Java Anon Proxy (JAP) research project (Berthold *et al.*, 2000a) at TU Dresden is a commercial service implementing cascading low-latency mixes. They claim higher security (and anonymity) by using external operated services in their cascade of mixes. The traffic between the user and the mixes also contains dummy traffic, or padded data, which will make the correlation of traffic harder.

Figure 5.4 Cascading proxies.

---

6 JonDo is a proxy client in the system JonDoNym, a commercial service running at https://anonymous-proxy-servers.net

Since this is a commercial service, a police investigation will be able to send a request for traffic monitoring to get information about the users, but this court order must be submitted to each of the operators of the mixes in the chain, which may be located in different jurisdictions. If these operators do not keep logs of high details and long duration or have been legally ordered to do so, this is unlikely to be of any use in an investigation of earlier communication traffic.

But even here, it may be possible to perform a traffic correlation attack. For example, if an attacker can observe traffic coming into the mixer cascade, e.g., from a suspected user, and the exit traffic from the last mixer in the chain. But even if the principle of this attack is known, we do not currently know of any successful designs proving such an attack on this kind of network.

5.2.5 Anonymity Networks

The popular free anonymity service available, Tor, uses a network of participating nodes to mix traffic, and we call this set of cooperating nodes an *anonymity network*. But there are several types of anonymity networks and many implementations and proposed services of each type. In order to create some structure of these anonymity networks, we have divided them up into three main categories:

— Broadcast protocols
— DC networks
— Source-rewriting networks

5.2.5.1 Broadcast Protocols

Broadcast protocols are quite simple to understand. You have a group of communicating nodes, and you broadcast the same information to all potential receivers. By having all potential senders sending encrypted content in a synchronized scheme, and where all potential receivers receive all information from all senders, we realize it will be really hard to determine who communicates with whom. But this comes with the cost of an enormous amount of unnecessary network traffic. Often these types of protocols use constant rate transmissions to achieve this.

Peer-to-Peer Personal Privacy Protocol, P5 (Sherwood *et al.*, 2002), is a broadcast protocol that tries to optimize network performance by creating broadcast groups in a hierarchy. This, of course, reduces the degree of anonymization that can be created through broadcasting to all and reduces network overhead to achieve higher throughput.

If a broadcast type of network is to be investigated, it will be near impossible to determine which parties communicate without breaking the encryption or having access to the end computers.

5.2.5.2 DC Networks

DC networks are named after David Chaum's paper introducing *the dining cryptographers' protocol, DC-net* (Chaum, 1988), where he proves *absolute anonymity* within a group of users that cooperates in sending messages. The principle is shown in Figure 5.5.

All users of the DC-net share a secret with at least two other users. This secret can be only one bit long. If each host transmits the *xor* of all shared bits, the sum of all transmitted bits will be divisible by two. If a host in a DC-net wants to send information, it

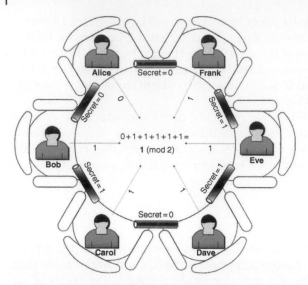

Figure 5.5 DC-net principle.

transmits the inverse of the actual value, and the sum will not be divisible by two, and a bit is sent from the group without compromising the origin of the message. An exercise is left for the reader to find out which of the participants are transmitting since the sum of all transmitted bits is not divisible by two.

Herbivore (Goel *et al.*, 2003) is an attempt to build upon the principles of DC-net to create a more efficient anonymization protocol, but without any resulting practical use. But optimization comes with a reduction of the anonymity set and thereby through a reduction of anonymity.

The DC-net's absolute anonymity within the set is an interesting but not a practical demonstration of anonymity.

5.2.5.3 Source-Rewriting Networks

Source-rewriting is used in all practical low-latency anonymity systems. The principle of source-rewriting is comparable to an anonymizing proxy since it involves changing the origin of the traffic. And the *network* term indicates that we have multiple such proxies working together. The last decade(s) advancements in low latency anonymizing technologies are mainly coming from the development of the *onion routing* principle (Goldschlag *et al.*, 1996; Reed *et al.*, 1996).

Onion Routing The term *onion routing* is named after the principle of having a message "surrounded by multiple layers of encryption," similarly to layers of an onion. When the message is sent over the network, another layer of encryption is either removed or added depending on the direction of the traffic and the implementation of the protocol.

The principle for onion routing is like advanced remailer mixers, but for the nodes to maintain relatively low latency on traffic, the network establishes a communication channel through a chosen set of nodes in the network. This communication channel is called *a circuit*. The selection of which nodes to use in the communication channel is an important research discussion, but randomness must be combined with nodes' ability to transport traffic (bandwidth), their stability (uptime history), location of the nodes (trust in local government and ISPs), owner of the nodes (known/unknown), trust in the nodes themselves, their administrators, the nodes' service providers, and more.

The principle of using public-key cryptography on all packets at all nodes will give a huge calculation overhead enforcing high loads of cryptography calculations on all nodes in the circuit. Therefore, onion routing uses only public-key cryptography to distribute so-called *session encryption keys* between the originator and every anonymizing node in the circuit. The procedure for setting up the circuit is shown in Figure 5.6, where the client, C, has chosen three nodes, N_1, N_2, and N_e, as circuit nodes for setting up a circuit. The client builds the packet from the bottom of the figure and up, similarly to "building an onion" from the inner layer by adding extra layers:

− The session key K_{CNe} shared with the last node N_e, is added to any command or message, CMD, it wants to perform at, or send to, N_e (e.g., an open Transmission Control Protocol (TCP) connection to a website), and this is encrypted with N_e's public key, Pu_{Ne}.
− Then, the identity of node N_e is added to the encrypted packet along with the session key K_{CN2} shared with node N_2, and the entire packet is encrypted with node N_2's public key, Pu_{N2}.

And finally, the identity of node N_2 is added to the encrypted packet along with the session key K_{CN1} shared with node N_1, and the entire packet is encrypted with node N_1's public key, Pu_{N1}.

Now the packet is ready to be transmitted from the client to Node N_1. Upon receiving the packet, node N_1 decrypts it with its private key, finds the session key to share with the client and use for this circuit, locates the identity of N_2, and forwards the remainder of the packet to N_2. Node N_2 performs the same procedure and retrieves the session key with the client and the identity of N_e, which have a similar task before finding the session key and the command CMD.

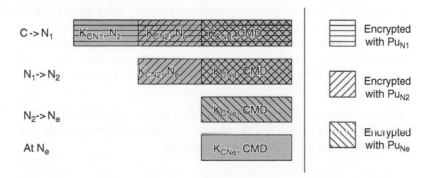

Figure 5.6 Setting up session encryption keys in onion routing (Øverlier, 2007).

CircuitID	K_{CN1} [K_{CN2} [K_{CN3} [Data or Command]]]

Figure 5.7 Small overhead when a circuit is established – the circuitID is changed, the packet is encrypted/decrypted one layer, but the length is not changed.

When the circuit is set up, the communication will travel a lot faster through the network and with far less overhead. Figure 5.7 demonstrates this principle where all we need is an identifier for the circuit, and no transmission of other information is needed for the transport of data or new commands. Sent data is encrypted with all symmetric session keys, which does not expand the ciphertext, and on the route through the network, one layer is decrypted at every node with the identifying information about a circuit, *a circuitID* is replaced with the next hop's circuitID.

Return traffic from the last node will look similar when arriving at the client. Each node will add one symmetric encryption layer to the data area, and the client will need to perform three (symmetric) decryptions to retrieve the data sent back to it from N_3. By using only known good algorithms for symmetric encryption and decryption, the security is high after the initial setup. But there are many challenges with predistributed keys, like replay attacks and no perfect forward secrecy (or anonymity).

5.2.5.4 Tor, The Onion Router

We will, in more detail, describe the functionality and use of the anonymizing network Tor (Dingledine *et al.*, 2004) as this is by far the most significant and widespread anonymizing service today with over two million simultaneous users[7]. Tor originated from the Onion Routing program[8] at the US Naval Research Laboratory in 2001–2002, and Tor is an improvement to the original onion router principle. Tor is still undergoing continuous improvements, upgrades, and changes, so please refer to the specifications of Tor for up-to-date information[9].

As of late 2022, Tor consists of approximately 7000 so-called *relay nodes*, or *server nodes*, run by volunteers in all parts of the world. A graph picturing the growth of the Tor network is shown in Figure 5.8. A list of all these publicly available nodes is stored and regularly updated in the *Tor Directory Authority*.

Let us try to enter the quite complicated details of the Tor network to see how it operates on the packet level. Figure 5.9 shows a network of seven relay nodes with a client seeking access and a normal Internet service running on a server. In addition, it is necessary to know that all nodes in the network use transport layer security, (TLS) tunnels as the underlying foundation for all other communication. This adds an extra layer of encryption to the principle of building the anonymous channels described here:

– The client downloads a list of the relay nodes, Tor Server 1-7000 (TS1, TS2, . . .TS7000), in the Tor network from the Tor directory mirrors, which provide this information to the world from the Tor directory authority.

---

7 https://torproject.org
8 https://www.onion-router.net
9 https://gitlab.torproject.org/tpo/core/torspec

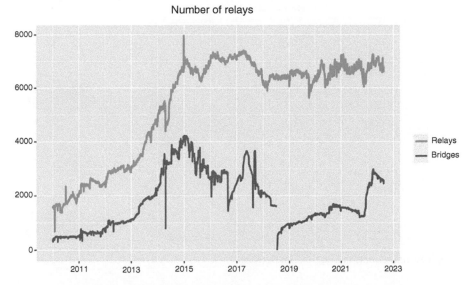

Figure 5.8 Number of nodes in the Tor network. (Tor metrics/Public Domain CC BY 3.0.)

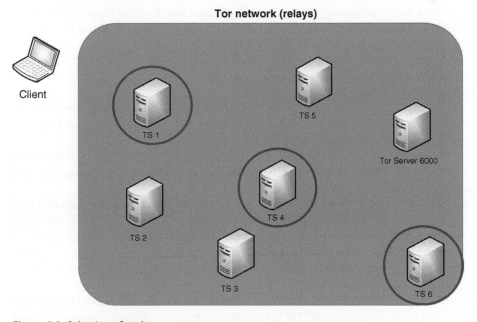

Figure 5.9 Selection of nodes.

Having this list of nodes, the client wishes to connect to the server and must carefully select a set of nodes to be used through the anonymizing network. In Figure 5.9, these are shown as TS1, TS4, and TS6. The first node chosen is the *entry node*, and the last node is the *exit node*, where the traffic exits the anonymizing nodes to communicate with "plain" services outside the anonymizing network. The nodes used between the entry and the exit nodes are called *middle nodes*, and usually, in Tor, there is only one middle node between the entry node and the exit node.

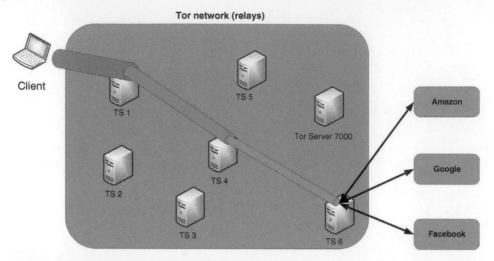

Figure 5.10 Layered tunnels through the Tor network.

The client sets up an encrypted tunnel to the entry node, TS1, using a secure key exchange and enabling perfect forward secrecy for the communication channel. This tunnel now acts like an anonymizing proxy, and all following traffic through this tunnel will look like it originates at the entry node. If the client is **not** a part of the Tor network nodes, TS1 will know it is being used as an entry node.

Through this tunnel, the client sets up an encrypted tunnel in the same way to the middle node, TS4. TS4 cannot tell whether the origin of the tunnel request is TS1 or any other computer on the Internet.

And through the new tunnel with TS4, the client sets up an encrypted tunnel to the exit node, TS6. TS6 can also not determine whether the origin of the tunnel request is TS4 or any other computer on the Internet.

Now the client can ask the exit node to make requests to any type of TCP service on the Internet and have the replies sent back through the tunnel to the client. The complete tunnel is shown in Figure 5.10.

For this to work seamlessly at the client, all communicating programs use this local Tor client as a proxy which handles all the communication with the network and makes all requests transparent for the user. The proxy service may also be a network Tor proxy, like a common NAT gateway, but then all local network traffic is unencrypted and without confidentiality and integrity protection between the client and this proxy, which may compromise anonymity.

If the client wants to connect to another Internet service, it may connect through the same tunnel to the exit node TS6. Or the client can set up an alternative tunnel through the network and have all communication with the new service leaving through another exit node.

There are a few additions that have been ignored above, which make the Tor network a little more complicated:

– All connections made from one Tor node to another are made using TLS, to achieve perfect forward secrecy. This means that there is an extra layer of encryption in every hop of the Tor circuit.

— Selecting the nodes randomly from the entire set of potential nodes makes the network more vulnerable. It will push much traffic on the low-bandwidth nodes and leave many high-bandwidth nodes with available resources. Tor adjusts for the nodes' bandwidth.

— The clients may voluntarily provide access to the Tor network through their Internet connection, now acting as what the Tor network refers to as *a bridge*. As the client's IP addresses are highly dynamic, rapidly changing, and not on any global list, these bridges will allow clients in restrictive environments (like censoring countries or companies) to connect through the Tor network. The number of bridges available is also shown in Figure 5.8. The major challenge of Bridges lie in the distribution of their IP addresses to the censored users without revealing them to the world. This is a research area of its own that we will not address here.

As we now have a simple understanding of how the Tor network routes traffic and sets up communication tunnels, we can look at some of the vulnerabilities.

5.2.5.5 Attacks and Vulnerabilities on the Tor Network

To perform correct and useful investigations on anonymity networks, we will have to understand the different attack scenarios. These are the most known attacks and a brief description of each.

Endpoint Traffic Correlation With most anonymizing networks, there is a vulnerability against traffic correlation at both ends. Suppose an attacker looks at the traffic going in and out of both the client and the server. In that case, the traffic patterns can, in most anonymizing networks/services, be correlated by relatively simple statistical analysis. This is a known attack that most anonymizing networks, including Tor, do not try to counter as it will require dummy traffic, leading to other potential problems and vulnerabilities.

Global Adversary The global adversary problem is an extended version of the endpoint traffic correlation challenge and is not a problem the Tor project aims to solve.

The Sybil Attack One of the more discussed attacks is the *Sybil attack* (Douceur, 2002) named from the Flora Schreiber novel "Sybil," about a woman who has multiple personality disorder. In the Sybil attack, a single attacker controls a large set of nodes in the Tor network. These nodes may cooperate by sharing traffic information and correlating and routing traffic to achieve easier anonymity compromise. The Sybil attack is also discussed in recent research papers (Winter *et al.*, 2016)

The Sybil attack simplifies the anonymizing part of the network to get closer to performing an endpoint correlation attack. The attack makes it more likely to know more about or control both the entry and the exit node. And this is also without having access to local traffic at neither the client nor the server (which most likely are in different jurisdictions).

Tor Node Availability One of the major problems for dissidents in using Tor is that it is relatively easy for the oppressing regime to stop all connections to the Tor servers, as an updated list of all the servers can always be downloaded from the Tor Directory service always. Banning all Tor nodes has been attempted in many countries already with various degrees of success (Paul, 2015), but the Tor network also allows another type of access outside the publicly listed server nodes described above as *bridges*, which are a lot harder to identify and therefore ban.

Entry Node Randomness – Timing Attack   A method to locate the originating client using limited resources in the network came through timing attacks (Levine *et al.*, 2004). When all nodes are equally likely to be selected as the first node in the anonymizing circuit, an attacker only needs one evil node in the network and a method to say whether he is a part of the circuit or not (Øverlier & Syverson, 2006b). If the client repeatedly makes connections where the attacker can evaluate if his node is a part of the circuit or not, there will be a list of originating IP addresses recorded. This list will have the IP address of random Tor-nodes when the attacker is an exit node or a middle node, but it will be the same IP address (namely the IP of the Tor initiator) whenever the attacker is the entry node. This means that the supposedly anonymized user's IP address will occur in approximately 1/3 of all verified usage where the attacker's node in determined to be used in the circuit.

A countermeasure towards this attack was implemented in Tor, called *guard nodes*. These are nodes with higher trust that the client uses as the entry nodes over longer periods of time, but these also have their issues (Øverlier & Syverson, 2006a). Guard nodes turn the above attack into an attack identifying the guard nodes themselves, and thus not the originating node.

Censorship-Resistant Publishing   Censorship-resistant publishing is the challenge of how to make information available to others without letting any entities have the possibility to censor and or shut down the service providing this information. For example, if a dissident published information critical of the regime on a local web service, the web service could be shut down easily. If the web service was in another country, it could be the target of legal or economic pressure, it could become the target of cyber-attacks, or it could be censored by the local regime trying to stop the information from reaching its own citizens. So, it was not only important to publish the information but also to make the publishing service resistant to attacks.

The three most important challenges in creating a censorship-resistant publishing solution were defined to be:

1) Disseminate information securely and anonymously.
2) Prevent non-authorized users from removing or changing information.
3) Prevent anyone from making the information unavailable.

5.2.5.6   Tor Hidden Services – Tor Onion Services
One such system supporting censorship-resistant publishing is the so-called *Tor hidden services* or *Tor onion services*. The design criteria for the hidden services were:

1) Be accessible from anywhere.
2) Resist distributed denial-of-service (DDoS) attacks.
3) Resist attacks from authorized users.
4) Resistance against physical attacks.
5) Enable censorship-resistant publishing.

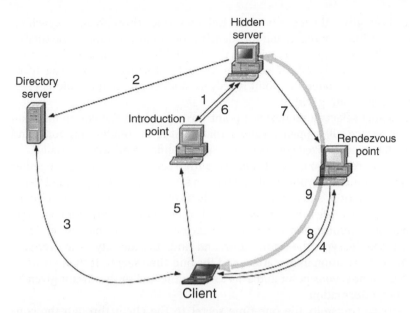

Figure 5.11 Setting up a hidden service connection (Øverlier, 2007).

The Tor onion services allow any service to be located anywhere on the Internet, just being a client or a server in the Tor network. In principle, it means to have a Tor client on the node where the to-be hidden service is located (which can be any type of TCP-based service, like web server, chat server, etc.). The Tor onion service allows all Tor users to connect to this service through Tor, but the service needs to be addressable. We will describe the principle of Tor onion services version 2, but the principle is mostly the same as in onion services version 3, which from 2021 is the only supported version in the Tor network.

Figure 5.11 shows the complex setup of Tor onion services. The connections made in each step are explained here:

1) To initialize the service, it needs to be identifiable, addressable, and reachable. For this, it builds a public/private key pair for the service. From the public key, the service derives a 16-character short name using base32-encoding, e.g., *abcdefghijklmnop*, which is used for identifying the hidden service as *abcdefghijklmnop.onion*. This short name will also assure the client that it connects to the correct service, but the readability and memorability of the service's name are now a challenge. In addition, the service chooses a set of *introduction points* from the Tor network. The service sets up a Tor tunnel to each introduction point and asks the introduction point to wait for authorized connections towards the service. The tunnel can be thought of as a plain Tor circuit where the exit node is used as an introduction point instead of exiting the network on the client's behalf. The introduction point(s) will be the hidden service's primary point(s) of contact.

2) The service creates *a service descriptor* containing the identity of the introduction point(s) together with their authorization code, time information, and the service's public key. This descriptor is then signed and published through the *Tor directory service*, where it is stored in a distributed hash table for later requests.

3) The client has been given the identity of the hidden service through an unspecified (but potentially off-line) source. Using the identity, the client downloads the service descriptor from the *directory service,* verifies that the public key identifies the service and that the descriptor is signed correctly, and extracts information about the introduction point(s). The introduction point is only used for contacting the hidden service, not for communicating with it.

4) The client randomly selects a rendezvous point through which the communication with the hidden service will happen, creates a Tor circuit to the rendezvous point, and asks it to listen for an authorized connection for the specified rendezvous circuit.

5) The identity of the rendezvous point and the authorization information together with a one-time secret is encrypted with the hidden service's public key, added to the authorization information, and sent to the introduction point through a Tor circuit.

6) The introduction point verifies the connection and authorization information and forwards the (encrypted) information about the rendezvous point to the hidden service.

7) The hidden service decrypts the information and finds the identity of the rendezvous point, the authorization information, and the one-time secret. It then creates a Tor circuit to the rendezvous point and asks to be connected to the circuit given by the authorization information.

8) The hidden service transmits the one-time secret to the client through the connected circuits to confirm that the resulting tunnel's endpoint is the hidden service.

9) After this, the tunnel lets the client talk to the hidden service, and no node between the client and the hidden service can tell anything about who is talking to whom. Both tunnels to the rendezvous point are through Tor circuits, as shown in Figure 5.12. So, if you think the communication with hidden services is slow, you are absolutely correct as it involves the encryption or decryption on six Tor nodes between the client and the hidden service.

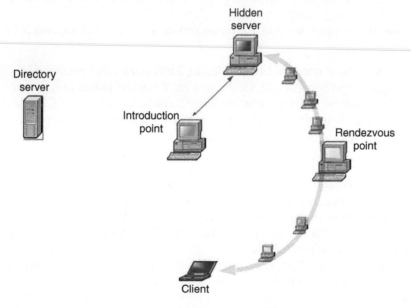

Figure 5.12 The resulting hidden service connection (Øverlier, 2007).

So, the directory service will contain all the hidden services that choose to publish their service descriptors. With version 2 of the hidden service descriptors, these are time-limited and made harder to link to earlier descriptors from the same service. Note that the client may still get the hidden service contact information in multiple ways, including off-line distribution, just like with onion addresses. An introduction point will only know the identity of the service if it is able to locate the service descriptors and identify that it is being used inside.

But all the communication with the hidden service *could* go through the introduction point and not through this complicated rendezvous setup. So why do we not use an introduction point not used for all the communication towards the hidden service? The main reason is that this might make the introduction point responsible or liable for the specific hidden service content. A rendezvous point *never* knows what it is connecting. It only shuffles data from one Tor circuit to another.

For more information regarding censorship and censorship-resistance on the Internet we recommend to check out the paper collection at CensorBib[10].

Tor Onion Services v3 The important changes for version three[11] of the Tor onion services are:

— Using elliptic curve cryptography to make key lengths shorter. Better encryption and hashing algorithms and more secure key lengths.
— Harder to link the time-limited service descriptors to the same service.
— Protected service descriptors.
— Introduces authentication tokens for access to the hidden service.
— Harder to make duplicate names. The identifying string (base32-encoded) from the entire public key, version byte, and checksum is increased to 56 characters (260 bits), making it even harder to duplicate (and remember).

5.2.6   Recent Live Messaging and Voice Communication

We will not address plain social media networks communication here as these normally do not allow anonymous usage. Most new communication platforms like Skype, Facebook, Twitter, LinkedIn, and Snapchat comply with law enforcement to assist in identifying their users, and they will be able to cooperate with investigators through normal legal procedures. We do not define these as anonymous communication platforms, even if there are many fake accounts and pseudonyms in these networks. And even then, international law and multiple jurisdictions have challenges of their own.

5.3   Anonymity Investigations

Performing forensics on anonymity systems is an extremely complicated and difficult task. We will address the challenges with a technical view, and without considering all the challenges of obtaining logs, IP address to physical address problems, involving multiple jurisdictions, and the coordinated efforts from many countries on the legal side.

---

10 https://censorbib.nymity.ch/
11 https://gitweb.torproject.org/torspec.git/tree/rend-spec-v3.txt

5.3.1   Digital Forensics and Anonymous Communication

We will look at three different situations and the challenges of digital forensics in the different scenarios. This section will try to list some of the investigation potentials, but several more potential and theoretical methods of attacks on anonymity systems can be found in the research community, e.g., at the paper database AnonBib[12].

The most realistic scenarios of anonymizing services are:

— Internet proxies with encrypted tunnels to proxy,
— Tor client usage for network access, and
— Tor client or server for serving hidden services.

Doing investigations requires access to data. We will reduce the challenge to address three forensic situations:

1) Using local logs and seized, powered off computers.
2) Using network logs and network access without (suspect) endpoint access.
3) The challenges and potential of live forensics and investigations.

5.3.2   Local Logs

Access to local logs can give useful information in many cases.

An Internet proxy service will have all its clients use encrypted tunnels (Virtual Private Networks (VPNs)) into one or more of its proxy nodes. Each node supporting the proxy service will mix all the clients into the node and make requests to the network (some Internet service) on their behalf, making it hard to see which client is connecting to which Internet service.

The anonymizing proxy running on a client is never supposed to log its activities. But depending on the tools using the anonymous services, the tools themselves may log the activities. Anonymized web services can be accessed through a specialized web browser, like the Tor Browser or a normal web browser with a proxy solution, which also will hide the user's IP address. The Tor Browser is usually tuned to avoid all logging and force all Domain Name System (DNS) lookup through Tor, but a normal browser may both store cookies, usernames and passwords, history, and more even if using an anonymizing service. Browser setup is significant, and there is a potential for user misconfiguration.

If a seized computer is one of the Tor server nodes, the file system will likely give access to the server's private key. The file system or the key itself may be encrypted, but the server must have access to the decrypted private key before it can rejoin the anonymizing network so it may be stored in clear text or decrypted through a startup script on the file system.

If a client is also hosting a hidden service, the hidden service's name and private key will be available in the configuration files with the same challenges as above. Access to the service's private key will enable live investigations of the hidden service's users.

5.3.3   Network Logs

With the enormous amount of traffic in today's networks, logs are only likely if there is some suspicion through the usage of a specific node, and logs, therefore, can be requested. Logs can be made of traffic to and from (1) the client (including performed at the entry

---

12 https://www.freehaven.net/anonbib

node of the anonymizing service), (2) an anonymizing server node, or (3) the accessed service (including performed at the exit node of the anonymizing service). So, we assume the logs consist of all relevant traffic in and out of the specific computer.

With logs from both the client and the accessed service in the same time period, suspicion of usage can be correlated and confirmed through timing and data traffic comparison (Øverlier & Syverson, 2006b). This is also correct for both plain network services and for traffic to and from hidden services.

5.3.4 Live Forensics and Investigations

Live forensics may sound like an oxymoron, but we try to separate live forensics and live investigations. If you are trying to locate users and servers of still active suspects, we can call it an investigation, but if you need to get transaction logs, user logs, activity logs, etc. from live systems after the evidence (supposedly) is located on servers where you have no physical control, we can call it live forensics.

Note! Breaking into suspected computers is considered out of scope for this chapter which only looks at de-anonymizing. All systems have vulnerabilities and abusing/attacking the software implementations for access to computers, but this is another area of research.

There are many methods of trying to de-anonymize and or investigate a live system, and we will try to cover some of them here. Some have been mentioned in the introduction to the different systems.

5.3.4.1 Timing Attacks
Timing attacks are mentioned above. They are highly efficient if there is no dummy traffic involved. With dummy traffic, the challenge becomes harder but remember that stopping an endpoint from communicating will also shut down dummy traffic which also will reflect on the rest of the anonymizing communication. Therefore, dummy traffic is not the complete solution to counter timing attacks.

5.3.4.2 Sybil Attack
The Sybil attack is also mentioned above. It is like fake social media accounts trying to boost its influence with more power than it is supposed to have (compared to its real size). Controlling large portions of an anonymizing service like Tor will make the attacker see more traffic and have more options in traffic analysis and traffic flow analysis.

5.3.4.3 Browser Fingerprinting
Web browsers give away information when used. Some browsers can be identified based on the browser type, libraries used in the browser, the operating system of computer, the size of the screen, fonts installed, languages supported, and so on (Abbott *et al.*, 2007). There have been several research papers explaining these vulnerabilities and how they can be used to identify or reduce the anonymity set of users (Hintz, 2002).

5.3.4.4 Language Analysis
Even if users can be anonymous, the users do not necessarily change the way they express themselves when talking, writing, typing, etc. This is why typing can be used for

the authentication of users. Language in texts produced or spoken may also be unique. Language analysis is also a separate area of research not addressed here.

5.3.4.5 Uptime Analysis

If a user can be traced to only be active in specific periods, e.g., through activity postings in the fake/anonymous social media account, the time periods active could be correlated to uptime periods of anonymous activities and/or participating anonymous network nodes. If a hidden service is only available when a specific server is up and running, this may be a coincidence once or twice, but not if correlation continues over longer periods of time.

5.3.4.6 Hidden Service Analysis

A hidden service may be open for all to access, like with the famous Silk Road service (Dingledine, 2013). But web site analysis of the service and all its pages and software may force the hidden service to give away information it was not supposed to give away, such as actual IP addresses, local usernames, compromising error messages, and local configurations.

5.3.4.7 Introduction Point Analysis

A hidden service can be effectively stopped if all introduction points are stopped from forwarding information requesting to connect to the hidden service. For the first versions of hidden services, it was relatively easy for an attacker to locate the introduction points for hidden services, and these were not frequently rotated. Now frequent rotation and hiding the hidden service name from the directory service have made this a lot harder.

5.3.4.8 Intersection Attacks and Entry Guard Mapping

The timing attack described in Øverlier and Syverson (2006b) will still work even after the guard nodes were introduced. But the statistical attack now only gives away the entry node and not the hidden service itself. Though, if you can map the entry node over some longer period and log the connections made to and from it, this will significantly reduce the anonymity set of the hidden service. If you later observe that the hidden service uses another entry node and you have access to log the connections made here, the intersection of these logs may be small enough to reveal the hidden service's location. But remember, this requires access to traffic logs at both suspected entry nodes.

5.4   Summary

In this chapter, we have looked at anonymity, how it has originated, what it is used for, and the usage in digital systems. We have given an overview of different types of anonymous communication protocols with a focus on low latency anonymity networks and the Tor network in more detail. We have also described how censorship-resistant publishing can work and how Tor's hidden services works as a tool for this. In the end, we have addressed the different types of attacks on anonymity networks and some thoughts on whether these attacks can be helpful in digital investigations.

5.5   Exercises

1   Explain the differences between privacy and anonymity. Can we have one without the other? Why (not)?

2   Use the list of legal and necessary uses of anonymity. Write down some legitimate examples from the bullet points and discuss what could be the implications if the ability to be anonymous was removed. Think about both positive and negative implications for society.

3   If each node in a high latency mixer chain of three mixes (from a set of 10 potential mixers) outputs from 50 messages at a time, what is then the probability for a message leaving N_3 to be the one you saw coming out from S?

4   Compare the vulnerabilities of cascading mixes and the Tor anonymity network from a user's viewpoint. Where does the user put his trust?

5   Which nodes in a Tor circuit are the most important? Why? (Trick question!)

6   Let us assume that one Tor server operator controls over 50% of the network and that all nodes contribute equally to traffic mixing. What is the probability of the operator controlling both the entry and the exit node?

7   Explain in your own words why the exit node does not know the IP address of the entry node in a Tor network.

8   Where is it most likely to find log files on a Tor server?

9   If suddenly 2000 new Tor nodes appear in the Tor server network. What kind of attacks can you assume are attempted?

10   What are the most likely reasons for identifying (getting the IP address of) Tor hidden services/servers?

6

Internet of Things Investigations

Jens-Petter Sandvik[1,2,]*

[1] *Department of Information Security and Communication Technology, Faculty of Information Technology and Electrical Engineering, Norwegian University of Science and Technology (NTNU), Trondheim, Norway*
[2] *Digital Forensics, National Cybercrime Center/NC3, National Criminal Investigation Service/Kripos, Oslo, Norway*

With the promise of billions of interconnected Internet of Things (IoT) devices, we can be fairly certain that digital investigations will involve systems such as personal area networks, wearable devices, voice assistants, connected vehicles, or other Internet-connected technologies. Estimates of the number of IoT devices have varied by orders of magnitude. In 2016, Intel estimated 200 billion devices by 2020, and Gartner estimated about 20 billion in the same year (Hung, 2017; Intel, 2018). An updated estimate by Cisco estimates the number of connected devices to be 29 billion by 2023 (Cisco, 2020).

No matter the number of devices, the future of digital systems tends to increase interconnectedness, storage, processing capacity, and the miniaturization of components. All things that can lead to new ways of using data systems and subsequently gives rise to new crime opportunities. As more and more IoT systems populate our surroundings, we will also see more and more evidence and investigations involving such systems. In this chapter, we explore some of the new investigation challenges that are introduced by IoT.

6.1 Introduction

As we saw in Chapter 2, the investigation process is our approach to reveal the who, what, where, when, why, and how about an incident. This process helps us minimize the cognitive biases that the human mind tends to produce. In addition, the chapter described the importance of hypotheses in an investigation and how one should not only consider competing hypotheses but also attempt to falsify each hypothesis. An investigation concerning IoT systems can be technical, and we might need to understand some of the fundamental technical details in order to be able to fully utilize the evidence from these systems. An IoT system can be involved in a crime in many ways,

* Jens-Petter Sandvik is a Ph.D. student in Digital Forensics at NTNU and a Senior Engineer at the Norwegian Criminal Investigation Service (Kripos). Email: jens.p.sandvik@ntnu.no

Cyber Investigations: A Research Based Introduction for Advanced Studies, First Edition. Edited by André Årnes.
© 2023 John Wiley & Sons Ltd. Published 2023 by John Wiley & Sons Ltd.

and one categorization of cybercrime that is useful for IoT systems can be paraphrased into three categories (Politiet, 2019):

1) The IoT system as a location of evidence.
2) The IoT system as a tool for a crime.
3) The IoT system as a target for a crime.

Of course, if the system is a target for the crime, it will usually also be a tool and an evidence location, but the opposite is not true. An IoT system can store video images of a suspect passing by as an example and be neither a target nor a tool for the committed crime. The categories span from pure digital forensics to investigations, where the collection of evidence from a location is the forensic aspect of an investigation, the tool usage is more related to both an investigative and forensic component, and the target of a crime has more to do with investigations specifically.

As an example of another type of category for digital evidence, we can look at the categories used by the European Convention on Cybercrime. This convention categorizes cybercrime into four areas (Council of Europe, 2001):

1) Offenses against the confidentiality, integrity, and availability of computer data and systems
2) Computer-related offenses
3) Content-related offenses
4) Offenses related to infringements of copyright and related rights

Items 1 and 4 are self-explanatory, and the former covers what previously was named as "system as a target for a crime," and the latter can also be named as "intellectual property crimes." As defined in the convention, the term *computer-related offense* means crimes where computers have been an integral part of the crime, and *content-related crimes* is a term reserved for crimes related to child exploitation material. More about the legal aspects and conventions governing cybercrime can be found, e.g., in Sunde (2018).

In this chapter, we will use the *Integrated Cyber Investigation Process* introduced in Chapter 2 as a basis for analyzing the challenges and opportunities that IoT introduces to digital investigations.

6.2 What Is IoT?

In order to understand the challenges posed by IoT and the opportunities offered, we must take a dive into what IoT is and how it works. Some of the more interesting technical details will be covered to a degree further than necessary for performing an investigation in IoT environments. This will help in getting a better understanding of the potential information available for the investigators.

6.2.1 A (Very) Short and Incomplete History

IoT is a paradigm that was used as far back as 1999 by Kevin Ashton to describe how things labeled with radio-frequency identification (RFID) tags could be read by Internet-connected devices and be tracked around the world (Ashton, 2009). IoT was thus a system that was, to a large degree, based on tracking things and processing information based on the location of the things. A typical application area for this was, e.g., logistics, where the transportation and storage of things could be optimized.

Later, the concept of actuator networks and Wireless Sensor Networks (WSN) was quickly adopted under the IoT umbrella, as the concept of things sensing and acting upon the environment was fitting into the concept of IoT and could be viewed as an enabling technology for IoT (Atzori *et al.*, 2014). A WSN is a network of sensors that typically operates in a network structure called a *mesh*, where each sensor connects to its neighbors, thereby creating a connection to the gateway of the network from a peripheral sensor without a direct connection.

IETF proposed IPv6 over Low-power Wireless Personal Area Network (6LoWPAN) as a Request For Comments (RFC) in 2007, which is a protocol for encapsulating the relatively huge IPv6 packets into the relatively small IEEE 802.15.4 frames (Montenegro *et al.*, 2007). Together with 6LoWPAN, IETF also described the IPv6 Routing Protocol for Low-power (RPL) and lossy networks as a routing protocol for mesh networks in RFC 6550, which is from 2012 (Winter *et al.*, 2012).

In the mid-2000s, there were many efforts in creating middleware for IoT based on service-oriented architecture (SOA) (Guinard *et al.*, 2010; Jammes & Smit, 2005; Spiess *et al.*, 2009). This is a way of designing a computer system focusing on the services offered and how to discover them. The thought was that each thing or group of things could offer some service, e.g. a sensor reading or a motor acting on a request. This opened the way for thinking of the things as a *service provider* for other applications, thereby letting one "thing" be used in many other applications. It meant that the services provided were not bound to any particular application, and new applications using the things could be easily implemented.

With interest in SOA came the problem with *service discovery* in IoT systems: How can all the services offered by the devices be detected by the clients? Various middleware solutions have different ways of discovering resources. Rathod and Chowdhary reviewed several middleware solutions for IoT and found that the majority only had local discovery (Rathod & Chowdhary, 2018). They also describe that global resource discovery is an open research issue.

The interest in IoT, both among researchers and the general public, has increased considerably over the last few years, as Figure 6.1 shows. The graph shows the number of published papers in the Scopus database and the Google trends statistics related to the term "IoT," and since 2010, there has been a clear and considerable increase of interest in the subject.

As the popularity of IoT has increased, new trends and research directions within IoT have emerged. Some of the new and specialized areas introduced new terms, such as *Web of Things* for the concept of web services-based communication between things, services, and users. We will come back to this term in Section 6.2.4. Other topics include the *Internet of everything*, where not only the things are in focus, but also the users, data, and processes are included in the model (Jara *et al.*, 2013). This term was coined in 2013 and gained popularity in 2014 and onward. There is also a recent interest in combining IoT with blockchain technology, from the introduction of blockchains focused on microtransactions between IoT devices such as IOTA (Popov, 2018), to the storage of information and sensor data in blockchains (Reyna *et al.*, 2018).

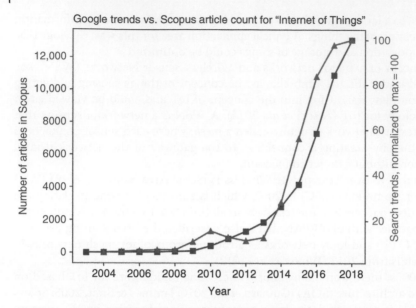

Figure 6.1 "Internet of Things" trends. The number of papers pr. Year in Scopus database on the left, and Google trends pr. Year, normalized to maximum = 100, on the right.

6.2.2 Application Areas

There are many application areas that have emerged under the IoT umbrella. This is by no means an exhaustive list of such areas, but a general overview to help understand the variety of fields that are covered by the term "IoT."

One of the first application areas people think of when asked about IoT is *smart homes* and home automation applications.[1] There are many products today that promise a more secure or convenient home. There are systems for adjusting lights and heat at home, online coffee makers, alarm and security systems, surveillance cameras, door locks, voice control or voice assistants, etc. Many of these systems are connected to a cloud service as the central control hub, and they are often controlled from a companion app on a mobile phone. At the time of writing, smart home systems are still expensive and are mostly used by people interested in technology and *early adopters*. If technology can help people save time or ease their daily chores, it will gain a wider acceptance. Figure 6.2 shows some of the devices that can be found in a smart home.

Intelligent transportation systems (ITS) and *smart vehicles* are application areas that focus on optimizing transportation challenges. By connecting cars to the Internet, they can receive over-the-air firmware updates, get information about the traffic situation, or automatically call for help in case of an accident. One of the challenges with connected vehicles is that the connection between the vehicle and the infrastructure is often interrupted as the vehicle is in areas without coverage or with a low-bandwidth

---

1 Many people make jokes about the inability to make food when the fridge will not talk to the stove, which points to the tendency of the current marketed IoT products to add complexity without a perceptible advantage.

Figure 6.2 Examples of smart home devices. From the upper left, we see an Internet-connected figure for kids, a Google smart home speaker, an Internet-connected fire alarm, and an Internet-connected baby monitor (Photos copyright Jens-Petter Sandvik).

connection. Inter-vehicle communication is, therefore, something that can resolve some of these challenges and even let vehicles share data without an infrastructure connection. A Vehicular Ad-hoc Network (VANET) is a way for vehicles to communicate by setting up a network between themselves that can route packets either between themselves or to an external network (He *et al.*, 2014). Example 6.1 discusses an attack against an Internet-connected vehicle.

Example 6.1: Jeep Cherokee hack

At the BlackHat US 2015 and DefCon23 conferences, researchers Miller and Valasek presented how they could remotely kill the engine of a Jeep Cherokee (Miller &Valasek, 2015). They were able to contact the vehicle through the cellular network by using the exposed DBus interface. The vehicle had two CAN-buses, one for the physical system, CAN-C, and one for the multimedia system, CAN-IHS, that were supposed to be air-gapped. Each of these buses had several electronic control units connected, and the CAN-C bus is the one that has the control units for steering, braking, and so on. Bluetooth, air condition, and so on, are connected to the CAN-IHS. One device was attached to both buses, and they could therefore access the CAN-C bus through this device and manipulate the signals sent on this bus.

(Continued)

> **Example 6.1: (Continued)**
>
> From a forensic perspective, an accident happening from a hack like this might be hard to detect unless there is an obvious reason to suspect it. The traces can be hard to find, as there was no patching of the system needed to perform this attack. This means that there are no obvious traces in the firmware or flash memory that can be acquired. There may be data on the behavior of the vehicle, but not necessarily of the commands sent. This means that while lacking direct evidence of the commands, we must infer the commands from the recorded behavior. The radio module might have traces on the external connections, but as the data in most cases only are found in RAM, diagnostic tools might not be able to access these traces.

Another application area is the *smart infrastructure* or *smart cities*. This is a term used for Internet-connected city-wide infrastructures. Some of the infrastructures that have been considered are road status and traffic monitoring, monitoring and activation of park watering systems, real-time parking space information, monitors for the structural health of buildings, just to name a few. One example of smart infrastructure is the testbed running in the Spanish city Santander, called SmartSantander (Sanchez *et al.*, 2014).

> **Example 6.2: SmartSantander Smart City**
>
> SmartSantander was a IoT testbed implemented in the Spanish city Santander. This project studied how IoT could be implemented and managed in a smart city environment and how the acceptance of smart city technology among the general population. The key technology used was WSNs for environmental monitoring like noise and air quality, but it also included RFID tags for augmented reality (AR) applications and sensors for parking spots (Moreno *et al.*, 2015).

Factories and industries have also embraced the promises of IoT and started researching and implementing IoT systems for their control systems. This is often referred to as *Industrial Internet of Things* (IIoT) or *Industry 4.0*. Factories and other industries have had industrial control systems (ICS) for a long time, but one of the challenges in transitioning to Internet-connected control systems has been the necessity for guaranteed security to their systems, as exposing fragile systems to the Internet poses a completely new risk to the security of the system. Another interesting application is *cloud manufacturing*, or what can be considered manufacturing as a service, where the manufacturing resources are shared and automatically assigned to the particular tasks (Tao *et al.*, 2011).

The first application areas for IoT with its history of Internet-connected tracking of RFID are probably *smart logistics* and *asset tracking*. The ability to always know where each asset is in real-time and dynamically optimize the usage or the route of assets can save a considerable number of resources.

One of the areas where IoT has been researched heavily is within *agriculture*. This is a field that historically has developed many automation methods to help maximize the yield and to minimize resource and pesticide usage. Monitoring crop fields with sensors for soil moisture, temperature, and lighting conditions are helping the farmers optimize their yield. Research in autonomous vehicles for harvesting or precision pesticide treatments is ongoing, and we will probably see more of this in the future (Bazán-Vera *et al.*, 2017).

Example 6.3: Fish farming cases

Fish farming is a big industry in Norway, and IoT technology can help decrease costs, increase efficiency, and improve the fish population health by monitoring the environment in the fish cages and subsequent analysis of the data (Parra *et al.*, 2018).

There have also been criminal cases filed against fish farms in Norway, and digital evidence from the operation of the farm can be used as evidence in cases that may span both environmental, animal welfare, and economic crimes in addition to cybercrimes.

Environmental monitoring is also a field where the potential for collecting and compiling data from resource-constrained, Internet-connected sensors promises a huge improvement over existing methods of gathering environmental data. Sites can be monitored in real-time, and events that can lead to natural disasters can be detected and the risks mitigated (Lim, 2015).

There is ongoing research on the use of Internet-connected medical devices and other IoT devices for use by health professionals. *E-Health* is used as a term used for electronic healthcare, and the term is not necessarily used exclusively for IoT-related research but also for the digitalization of archives (Oh *et al.*, 2005) or other digital improvements to health care. Mobile health, or *m-Health*, refers to health care services that can be provided independently of location (Silva *et al.*, 2015). The health personnel might be given access remotely to the patient's current medical data in order to give more efficient and effective treatments. There are other concepts such as *ambient assisted living*, a concept for elderly care using connected devices to increase safety and living standards for the users and to better utilize healthcare resources (Dohr *et al.*, 2010).

Table 6.1 shows a summary of the application areas described here, together with examples of the areas.

Table 6.1 Some application areas of IoT.

Application area	Example
Smart homes	Security camera and fire alarm
Intelligent transportation systems	Connected cars
Smart infrastructure	Citywide information system
Industrial IoT	Factory automation
Smart logistics	Fleet management
Environmental monitoring	Flood warning
E-health	Mobile health technology

6.2.3 Models and Concepts

There are many different models and architectures that have been proposed for IoT. A typical model is based on the concept of a local network, a gateway, and a cloud service (see Figure 6.3(B) + (D).) The local network may be an ordinary Local Area Network (LAN), a mesh network like ZigBee, or other types of networks. In order to communicate over the Internet, a *gateway* will either translate each internal address from the particular addressing scheme to an IP address, so each node is globally addressable or just collect the data in the local network and act as a router or address translator for the local network.

The data from the nodes might then be stored and processed in the cloud, or a link to the resources provided by the nodes are stored in a resource directory, or services stored in a service directory (Katasonov *et al.*, 2008). The nodes on the local network are typically resource constrained, both in terms of processing power, memory, and electric power usage, therefore the network protocols and security measures must be lightweight and is not necessarily compatible with the TCP/IP network stack. A gateway is thus needed in order to route the data and transcode packages.

Another way of connecting things to the Internet is directly from the device and through an access network to either a local service or a cloud service. This is a typical setup for remote nodes that uses, e.g., a 5G network to connect to the Internet. Some of the current methods for direct connection are through mobile 5G networks like Long-Term Evolution (LTE) or a Low-Power Wide Area Network (LPWAN) like LoRaWAN, Narrow-Band IoT (NB-IoT), and SigFox (Figure 6.3(A).)

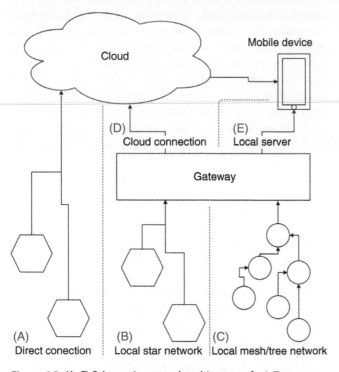

Figure 6.3 (A–E) Schematic network architectures for IoT systems.

Figure 6.3(E) shows another popular way of connecting an IoT system to the Internet, by letting the gateway act as or forward traffic to a local server where clients can connect.

A mesh network, like a WSN, is depicted in Figure 6.3(C) is a set of interconnected nodes that can route traffic between nodes that are not directly connected and where each connection can dynamically change. This setup typically must use a gateway in order to translate traffic between the mesh nodes and the IP side of the network.

While the IoT system can be connected to a cloud service, there might be requirements in the form of bandwidth limitations, energy usage, or latency between local nodes in the system. Much of the computation can therefore be performed closer to the edges of the system, and this is typically called *edge computing*. The term was originally used for web caches at the edges of content delivery networks but is today used in a more generic way (Ramaswamy *et al.*, 2007; Salman *et al.*, 2015).

One of the more interesting features of the edge computing concept is where the cloud storage and processing can be moved from a centralized infrastructure toward the edges of the system. The concept of the cloud coming closer to the edge has been called *fog computing* (Bonomi *et al.*, 2012). In this paradigm, the storage will be spread over one or more local "clouds" that are closer to the nodes, and data can be migrated from one of these local clouds to another if the locations of the nodes change, for example, if a connected vehicle moves from one area to another.

Definition 6.1: Edge computing

The computation of data in close proximity to where the data is generated and/or used, close to the edges of the system in contrast to computing in a cloud or a centralized server.

Definition 6.2: Fog computing

A specific implementation of edge computing, where local clouds or *cloudlets* are used. Data can be transferred between these local clouds if the data is needed in another location.

One of the terms often encountered in IoT is the concept of machine-to-machine (M2M) communication, where the machines communicate between themselves independent of human involvement. M2M is a broad concept, and while the communication technology might be the same, the M2M paradigm promises more communication that is not initiated by humans in the system. The initiation of communication is done autonomously by a machine, and the response is also autonomously handled by a machine. From an investigation point of view, this means that the traffic must be properly accounted for in order to decide whether it is a human initiating the traffic or a machine initiating an autonomous M2M session.

Definition 6.3: Machine-to-machine communication

The automatic communication between computing devices that is not intended for human interaction.

Cyber-physical systems (CPS) is a term used for describing systems that have both a computing-based component and a component acting on the physical world. Typically, there is a feedback loop from the physical part to the computer part and a feedback loop from the computer part to the physical part (Lee, 2008). For all practical purposes, we can view the IoT as a continuation of CPS and building upon and extending the CPS model.

Definition 6.4: Cyber-physical system

A system that senses data from and/ or acts upon both the physical world and cyberspace.

One of the most prevalent attributes of IoT is the sensing capabilities of its environment. WSN was already described as a concept in 1996 (Bult *et al.*, 1996), and in the mid-2000s, the research field gained popularity. The concept of a WSN is that sensors can be distributed over a wide area, configured as a mesh network, and transmit the readings through the intermediate nodes to a gateway that is connected to the Internet or to another system that can process the readings. The sensor nodes may have strict limitations on their power usage, physical size, processing power, or storage size. The advantage of WSNs is that the sensors can cover a big area, where each transmitter only needs the power to transmit to its nearest neighbors, and that it is inexpensive to deploy when the infrastructure for communicating with the devices are built into them. Interference between the nodes may be minimized because the sending effect in each node is so small that it will not affect nodes further away.

Many inexpensive sensors can also be as precise as one expensive sensor. We are not going into the mathematical details here but rather discuss the intuitive observations. It seems rather obvious that if we take the average measurement from two devices, it gives me a slightly better estimate than the readings from only one device, and if we average the measurements from three devices, it is even better than from two. On the other hand, it also seems obvious that the one extra measurement from one to two devices increases the precision much more than the one extra measurement from one million to one million and one devices. This is, in fact, true, and the increase in precision for measurements is proportional to $\dfrac{1}{\sqrt{N}}$, where N is the number of measurements. Example 6.4 shows a numerical example of this. For a more detailed introduction to the mathematics of precision of measurements, please refer to a book on observational science like (Chromey, 2010).

> **Example 6.4: Precision of many imprecise nodes vs. one precise**
>
> If we assume that we have one very expensive and precise temperature sensor that can measure temperature with a standard deviation of 0.1°, and we also have some number of inexpensive sensors with a standard deviation of 1°, how many inexpensive sensors must we average in order to achieve the same precision as the expensive sensor? We can assume that there are no systematic errors affecting all the cheap sensors, that the average systematic error is on average 0, and that the errors in the readings are following a gaussian distribution, where it is more probable that the measurement is closer to the actual value than further away.
>
> If we use the proportional relationship between the precision and the number of measurements, $s = \dfrac{1}{\sqrt{N}}$, where s is the desired precision, this means that in order to increase the precision from 1 degree to $\dfrac{1}{10}$ degree, we need 100 measurements from the cheap sensors.

A digital twin is a term used in IIoT and is a digital representation of a device, complete with a model, data, metadata, and structure such that the twin can be used for simulations (Rosen *et al.*, 2015). The idea is that all data from a physical thing will be mirrored by the digital twin in, or close to, real-time. Decisions affecting a highly complex environment can be simulated, optimized, and planned before being sent to the actual device. A digital twin can also be used for automating production steps and go from automated systems to autonomous systems. The difference between these systems is that an automated system is where the production steps are planned prior to the start of the production, and an autonomous system is where the system itself makes decisions based on the unique characteristics of each item or environment.

One branch of research that is especially interesting from an investigation or forensic point of view and has been introduced with IoT is the concept of *social networks* between *things* (Atzori *et al.*, 2014). This concept is based on how *things* can decide the trust in an M2M communication network. There are different levels of trust assigned to how *close* a relationship there is between the things, and this closeness can be calculated from the closeness in device type, geographical closeness, the user's friends' devices, etc. This network will be dynamically updated and stored in a *Social IoT server*. This social IoT server will be interesting from an investigative standpoint, as it might hold evidence of interactions and trust relations between people and between people and devices.

6.2.4 Protocols

With the introduction of systems that have limitations to their processing power, energy consumption, and network bandwidth comes many new communication protocols for communicating between the devices themselves, the hubs in the network, and the Internet. While there are many competing standards and protocols, the most important protocol that ties everything together is Internet Protocol version 6 (IPv6). Another set of protocols that will be described is the Internet Engineering Task Force (IETF) stack of protocols for IoT.

The Eclipse Foundation has since 2015 conducted the annual *IoT developers survey* among IoT developers (Eclipse Foundation, 2018). This survey has shed some light on the technologies used in the respondents' IoT projects. As of 2019, the survey revealed that the top three communication protocols are HTTP (49% of participants use this protocol in their product/development projects), MQTT (42%), and WebSocket (26%). The three most used protocols for connectivity are still TCP/IP, Wi-Fi, and Ethernet. This means that the traditional protocols that we already know are still the most widely used.

As we will focus on the protocols used in IoT systems here, we start with IPv6, continue with a protocol for low power and lossy networks, and in the end, describe the MQTT protocol.

6.2.4.1 Internet Protocol Version 6

The first IPv6 specification was published already in 1995 as IETF RFC number 1883 and later updated in RFC 2460 in 1998. In 2017, IETF announced that IPv6 had become an Internet standard in the form of RFC 8200 (Deering & Hinden, 2017). IPv6 has a routable address space of $2^{128} = 3.4 \times 10^{38}$ addresses, as opposed to IPv4, which only has $2^{32} = 4.3 \times 10^9$ addresses, which is one of the reasons IPv6 has become one of the key protocols in IoT systems.

With longer addresses, the dotted notation used for IPv4 can be cumbersome to use. An IPv4 address might be 10.93.145.31 or written as hexadecimal bytes as 0a 5d 91 1f.[2] The IP address is 4 bytes written as decimal numbers between 0 and 255, separated by dots. The IPv6 address, on the other hand, is 128 bits long and is written as 8 16-bit words, using hexadecimal notation instead of decimal, separated by colons. The address also allows shortening a series of zero-words to a double colon. The IPv6 address 2001:db8::3 is, therefore, the address 2001:db8:0:0:0:0:0:3. This double colon can only be present in one location of the address in order to avoid any ambiguities as to how many zeros each shortening should contain.

When looking at IPv6 addresses, we can often see a slash followed by a number. This specifies an address prefix, and the slash tells us how many bits from the start of the address are used for the network prefix. This is also referred to as *Classless Inter-Domain Routing (CIDR) notation*. The address space used for documentation and examples is described as 2001:db8::/32, where the address range spans from 2001:db8:: to 2001:db8:ffff:ffff:ffff:ffff:ffff:ffff. 32 bits is two 16-bit words or 4 bytes that we see at the start of the address (2001,db8).

As with IPv4, there are some address ranges that are reserved. These are specified in RFC 6890, and an overview is shown in Table 6.2 (IANA, 2018). If these addresses come up in an investigation, they indicate a special address range is not necessarily publicly routable.

6.2.4.2 IETF WPAN Protocol Stack

A protocol stack is a way of splitting up a communication channel into different layers, where each layer has its own responsibility and functionality in order to send data from one program running on one device to another running on another device. Table 6.3

---

2 Or as an uint32_t: 0x0a5d911f in network order (big endian).

Table 6.2 IPv6 reserved addresses (Data from IANA (2018)).

Address	Description	RFC
::1/128	Loopback address	4291
::/128	Unspecified address	4291
64:ff9b::/96	IPv4-IPv6 translation	6052
64:ff9b:1::/48	IPv4-IPv6 translation	8215
::ffff:0:0/96	IPv4-mapped address	4291
100::/64	Discard-only address block	6666
2001::/23	IETF protocol assignments	2928
2001::/32	TEREDO	4380
2001:1::1/128	Port control protocol anycast	7723
2001:1::2/128	Traversal using relays around NAT anycast	8155
2001:2::/48	Benchmarking	5180
2001:3::/32	AMT	7450
2001:4:112::/48	AS112-v6	7535
2001:5::/32	EID space for LISP (managed by RIPE NCC)	7954
2001:10::/28	Deprecated (previous ORCHID)	4843
2001:20::/28	ORCHIDv2	7343
2001:db8::/32	Documentation	3849
2002::/16	6to4	3056
2620:4f:8000::/48	Direct delegation AS112 service	7534
fc00::/7	Unique-local	4193
fe80::/10	Linked-scoped unicast	4291

Table 6.3 OSI reference model for networking.

Layer	Name
7	Application layer
6	Presentation layer
5	Session layer
4	Transport layer
3	Network layer
2	Data link layer
1	Physical layer

shows the OSI model, which is often used as a reference model for all other such network protocol stacks. As this is a reference model, the real implementations do not always match up with the functionality described in each layer of the OSI model. The Internet, as we typically know it, uses (from the top) HTTP, TCP, IP, and Ethernet or Wi-Fi as the most common protocols. HTTP is an application-level protocol, but in the OSI model, it spans Application, Presentation, and most of the Session layer. TCP spans some of the Session layer and the Transport layer, IP covers the Link layer, and Wi-Fi (MAC and PHY) covers the Data link and Physical layers. Figure 6.4 shows how the protocols at various levels are packed into the protocol in the underlying layer for TCP/IP. Figure 6.5 shows the same for 6LoWPAN.

The IETF Wireless Personal Area Network (WPAN) protocol stack is a set of protocols building on top of each other in order to work in resource-constrained devices that are connected in a mesh network. A network that has limitations on power usage and is meant for personal area networks can be named a Low-power Wireless Personal Area Network (LoWPAN). A home network is typically set up in a configuration we can call a star network, where each device is connected to a central router. A mesh network, on

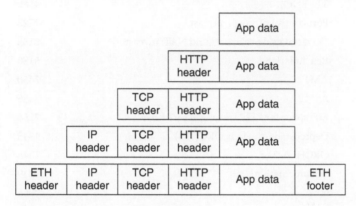

Figure 6.4 Protocol stacking in TCP/IP.

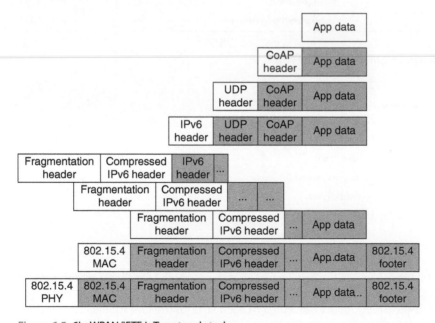

Figure 6.5 6LoWPAN/IETF IoT protocol stack.

the other hand, is a network where nodes are connected to each other, and traffic is relayed through the nodes in this mesh until it reaches the gateway.

The lowest level in the IETF protocol stack is the IEEE 802.15.4 physical and link-level protocols. These protocols define a WPAN, which is a network meant for resource-constrained radio interfaces and processing power. This defines the physical transmission of the radio signals, and the Media Access Control (MAC) layer defines the data frames. These data frames are not the same as Ethernet data frames but are at the same level in the protocol stack and have similar functionality.

6LoWPAN is a tunneling protocol that makes it possible to use IPv6 over LoWPAN as described by IEEE 802.15.4. Packets from the physical layer in 802.15.4 can be at most 127 bytes long, and excluding the overhead, this means 102 octets for payload. The media access layer content, which is 1/10th of the IPv6's required minimum MTU of 1280 octets, so 6LoWPAN works as a *fragmentation and reassembly adaption layer* (Montenegro *et al.*, 2007).

The next protocol in this stack is the IPv6 protocol, as described earlier. This layer also introduces a routing protocol called RPL. This protocol is the one that creates a route through a mesh network by creating tree-like traffic routing paths to the interconnected mesh network.

For the transport layer, User Datagram Protocol (UDP) and Internet Control Message Protocol (ICMP) are used. Transmission Control Protocol (TCP) is not used at all in this stack, which means that the higher-order protocols must implement retransmission methods if a lossless transmission of data is needed.

The application layer in the stack uses Constrained Application Protocol (CoAP). This is a protocol that, in many ways, is like HTTP but has some optimizations with regard to low-power and lossy network architectures: While HTTP is a text-based protocol where all the commands and return codes are text, CoAP is a *binary* protocol. This means that the protocol overhead is minimized, as text representations take considerably more bits to transfer than a binary representation. Another difference is that while HTTP is defined within a server-client architecture, where the HTTP server gets requests and sends the answer back to the client, CoAP lets both parties act as both servers and clients. CoAP also includes commands for retransmission of lost packets since the stack uses UDP as a transport layer instead of TCP. The earlier mentioned Figure 6.5 shows an overview of the IETF WPAN protocol stack.

In a network where all the nodes are connected by a link with all other reachable nodes, and where the link connections can change dynamically, it is necessary to have a way of routing traffic such that the traffic can dynamically be routed and without cycles in the network. The Routing Protocol for Low-power and Lossy Networks (RPL) is defined in RFC 6550 and creates a destination-oriented directed acyclical graph (DODAG) where packets can be routed from the nodes to the root of the graph or from the root to each of the nodes.

A DODAG is a tree with a root or a destination. This root is the gateway to the network and is the node that starts creating the tree structure. In order to create the tree, three ICMPv6 RPL packets are used. These are:

- DIS: DODAG Information Solicitation
- DIO: DODAG Information Object
- DAO: Destination Advertisement Object

DIS can be sent to request information from neighboring nodes, which will answer with a DIO. The DIO contains information about the rank of the sending node. The rank is information about how far the node is from the root of the tree, and if the node does not know other nodes in the network, the advertised rank will be very high. The receiving node can then select the node with the lowest rank in the neighborhood to connect to. This way, every node can create a tree structure towards and communicate with the gateway. DAO packets are sent to propagate information about the node upward in the tree. This way, packets can be routed both from the peripheral node to the root and from the root to the peripheral node. Figure 6.6 shows a DIO packet from the Cooja simulator that is a part of the ContikiOS project.

6.2.4.3 Message Queues

Message queues are a way of communicating between processes in a computer or between computers. A message queue is an *asynchronous communication channel*, which means the sender does not need to wait for any acknowledgments from the receiver, and the sender process does not even need to exist when the receiver reads the information (Stevens, 1999). Message queues were implemented as an interprocess communication (IPC) method for communication between processes in one computer system under System V. In this system, the kernel functioned as the message broker and received and stored the messages in a linked list for the receiver to get. Message queues have also been implemented over network connections and are now popular communication protocols for IoT devices.

MQ Telemetry Transport (MQTT): One of the more popular communication protocols for IoT devices is MQTT (Mqtt.org, n.d.). This is a protocol that was originally

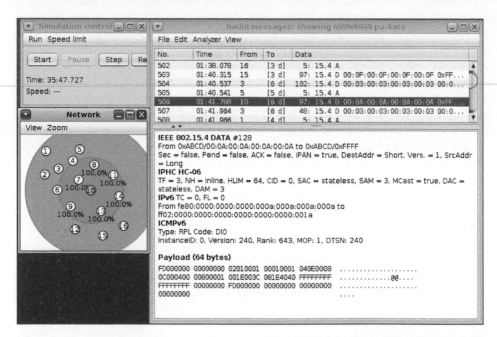

Figure 6.6 Cooja simulator showing a DIO packet. To the left is a diagram that shows the layout of the nodes.

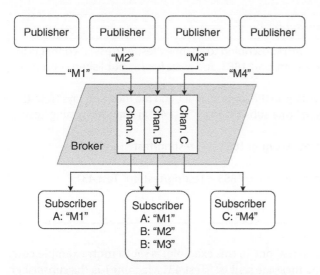

Figure 6.7 A Publish/Subscribe (or pub/sub) architecture.

developed in 1999 by Andy Stanford-Clark (IBM) and Arlen Nipper (Arcom). One of the reasons MQTT has become popular in IoT is that the protocol itself can easily be implemented in resource-constrained devices and that the publish-subscribe paradigm works well in typical IoT systems.

The publisher sends a message in a *topic* to a message broker or server. Subscribers connect to this server and subscribe to topics. Each time a message is sent in a topic, all subscribers to that topic will receive the message. Figure 6.7 shows how this architecture works. In this figure, the publishers each send one message ("M1" to "M4") to their channels ("A," "B," and "C"), and the subscribers receive each message from their respective subscribed channels. The message itself can be published as a retained message, which means that the broker keeps the message and sends it to each new subscriber. Messages without the retain-flag set will not be stored by the server and immediately be sent to the connected subscribers.

The forensic-aware student will probably ask, "what is meant by stored here? Where is it stored, and for how long?" The answer is: it depends on the broker. The open-source Mosquitto server has a configuration setting, *persistence*, that decides whether the data is stored only in memory or if it will be written to a database file. The default setting running without a configuration file is to store in memory only, but, e.g., an Ubuntu system will at installation time modify the default behavior and enable persistent storage in the configuration file /etc/mosquitto/mosquitto.conf.

The topics that the clients publish and subscribe to are not stored by the broker. To learn which topics are sent, we can either subscribe to topics using wildcards and collect posted messages this way, or we can analyze the incoming traffic to the broker, dump the running memory of the broker process, or examine/reverse engineer the various publisher nodes to see what they are publishing. For the example shown in Example 6.5, a wildcard topic subscription can be "testing/#".

A short example of the use of MQTT with the open-source implementation *mosquitto* is shown in Examples 6.5.

Example 6.5: MQTT publish and subscribe

Machine1, in this example, starts with subscribing to channel "testing/test", it connects to the MQTT broker at mqtt.example.com, port 1883, and waits for any published content in this channel.

Machine2 then publishes a message in the same channel, and the message says "Test 43". The result shows up at Machine1 (the subscriber) as "Test 43" in channel "testing/test":
Machine1
$ mosquitto_sub -h mqtt.example.com -p 1883 -v -t testing/test
Machine 2
$ mosquitto_pub -h mqtt.example.com -p 1883 -t testing/test -m "Test 43"
Machine 1
→ testing/test Test 43

We can see that the server (not a real one in this example) is set to mqtt.example.com, the topic is: "testing/test", and the message sent is: "Test 43". Machine1 is the subscriber, waiting for messages, and Machine2 is publishing the message, which in turn is received and displayed in Machine1. In a real IoT environment, the topic can, e.g., be "home/sensors/temp1" for temperature readings.

The server can be set up with Transport Layer Security (TLS) client and user authentication, but this is not enabled by default and must be configured by the system administrator. The log file will log the attached clients with both the IP address and the client's name. The messages themselves are not stored in the log file.

The broker, or server, can be on the local network but can also be hosted in the cloud or another network if it is reachable from both the publishers and subscribers.

6.2.4.4 Web of Things

Web of Things is an architecture which uses web protocols for communicating between the elements of the system. Representational State Transfer (REST) Application Programming Interfaces (APIs) are typically used for keeping the state of the resources, and JSON-LD is used for data transfer. JSON-LD is a data format for linked data that builds upon JSON (W3C JSON-LD Community Group, n.d.). Linked data refers to the use of Internationalized Resource Identifiers (IRIs) for referencing other resources, and an IRI is a resource identifier that can use the whole Unicode set as characters and allows for converting to a Uniform Resource Identifier (URI) (Duerst & Suignard, 2005).

Definition 6.5: Web of things

Web of Things is the paradigm that all things are using web protocols and technologies for communication, services, service discovery, etc. These web protocols are typically HTTP, WebSocket, JSON, etc.

World Wide Web Consortium (W3C) has published specifications for Web of Things (World Wide Web Consortium, n.d.). At the time of writing, this is a candidate recommendation that is meant to become an official recommendation from W3C.

The Web of Things architecture is based on four building blocks: The *WoT description of a Thing* is a description of metadata and interfaces to the things, and the *WoT scripting API* describes the JavaScript API for the things. The *WoT Binding Templates* provide guidelines on protocol bindings or how to make the various protocols used for IoT systems translate to the WoT protocols. The last building block is the *WoT Security and Privacy Guidelines* that, as the name implies, gives guidelines for how each of the other building blocks should consider the security and privacy issues.

While WoT typically uses HTTP(S) and Web Services (WS) as the application layer protocol, it can also use other protocols that have the needed capabilities, such as CoAP and MQTT.

6.3   IoT Investigations

An investigation often covers many leads and hypotheses, and some of them might include IoT systems. In this section, the focus is on the part of the investigation that involves IoT systems in one way or another. As mentioned in the introduction to this chapter, we can divide the crime into three overlapping segments, where the IoT system can be a location of evidence, a tool for the crime, and/or a target of the crime. For an investigation, we need to at least assess the nature of the investigation with regard to the IoT system.

Example 6.6: Mirai botnet

The Mirai botnet is a case where IoT systems can be seen as a tool for crime (Krebs, 2017). The malware targeted various Internet-connected Linux devices with default credentials selected among a list of 62 entries and was running on a variety of hardware architectures. ARM, ×86, SuperH, and MIPS were reported on VirusTotal in August 2016.

Most of the infected systems were IP cameras that were exposed on the Internet and where the default login credentials had not been changed. The malware was the only resident in RAM and relied on reinfection in case of a reboot (Antonakakis *et al.*, 2017). The source code for the malware was released online, and many IoT malware families after this are either based on Mirai or borrowed from the Mirai codebase. Later versions of Mirai used security vulnerabilities in the devices in addition to default credentials.

We can look at the malware from two perspectives: Various websites were the targets of Distributed Denial of Service (DDoS) attacks, and the tools for performing these DDoS attacks were Internet-connected surveillance cameras. If instead of focusing on the DDoS against websites and we focus on the break-in and installation of unwanted code in these IP cameras, the target of the crime was obviously the IP cameras. We should therefore be careful about the perspective of the crime and investigation so we get the full overview in order to properly understand the case.

Another way to think about this categorization is to think of it as the levels of trust we can have in the investigated systems. Data from an IoT system that has sensed some events in its environment and that has neither been accessed nor controlled by any of the persons involved in the incident can be assigned a higher level of trust than a system under the control of an involved person.

One of the challenges in investigating IoT systems is that the system is not necessarily well managed. In a smart house, the administrator of the systems is typically the owner of the house, who may not have the time or knowledge to properly manage the system. The system can therefore be malfunctioning or misconfigured, such as having wrong clock settings on a security camera. Another challenge is that devices that are managed by a third party. An alarm system can have a third party that manages the system, and the homeowner buys or rents the devices from this trusted third party. On the other hand, many systems can be well managed, and companies with both in-house expertise and economic incentives to their digital security will probably have well-managed IoT systems as well.

One interesting finding from the 2018 version of the Eclipse foundation's annual IoT developer survey was that when asked about what type of IoT data was stored, 61.9% of the respondents answered that they store time-series data, and 54.1% store log data (Eclipse Foundation, 2018). This is typical data that can be valuable in an investigation. Device information can be used for answering some hypotheses, but the data generated by the device as a response to external events is most often the data that can be used for strengthening or weakening an investigation hypothesis.

6.3.1 Types of Events Leading to Investigations

We have already seen an IoT crime in the Mirai botnet. The original code was created to be a tool for DDoS attacks and would infect IP cameras, routers, and digital video recorders (Kolias *et al.*, 2017). In this scenario, we can see at least two types of crime being committed. One is illegally breaking into the computer system, in this case, the IP cameras. The other type is the crimes, the IP cameras are commanded to perform against other machines, either breaking into them or performing DDoS attacks. An IoT botnet can have the same functionality as its computer counterpart. The installed malware can, e.g., collect information, render the system unusable, or act as an infrastructure for other crimes.

Example 6.7: Stuxnet

Stuxnet is a well-known malware that targeted centrifuges used for enriching uranium and changed their operation such that they misbehaved while reporting normal operation (Langner, 2011). Stuxnet was a highly sophisticated attack that tried to keep a low profile by limiting the number of infections and spread via USB sticks and local networks.

Another example of sabotage is the Triton malware that targeted Triconex Safety Instrumented System, a system that can perform emergency shutdowns in industrial control systems (Johnson *et al.*, 2017). This malware was designed to override the safety system and cause physical damage. Industrial IoT systems are especially exposed to industrial espionage, as they (by definition) are Internet-connected, thereby offering an easier attack vector that is harder to trace back to the adversary.

Of course, sabotage and DDoS botnets are not the only types of events leading to investigations. IoT systems often are implemented as resource-constrained devices and can be located in clandestine environments. As with many other computer systems, the devices might not necessarily have adequate security for the environment in which it operates, where they might be either unmanaged, managed in an insecure way, or contain new vulnerabilities. This means that while the device itself might not contain interesting data or functionality, it can be used as an attack vector to get a foothold into a network. Computers and servers on the internal network can be exposed to external threats via the IoT device. One such reported incident is shown in Example 6.8.

Example 6.8: Fishing for databases

In 2017, an American casino was the target of an intrusion where the high-roller database was accessed, and data exfiltrated to a server in Finland. The attack vector to get into the network was an Internet-connected thermostat in a fish tank in the lobby of the casino (Larson, 2017).

While not much information on this hack has been publicly revealed, we still can see the importance of looking at all connected parts of a network in order to find the original events and mechanisms in a cybercrime investigation.

Other types of crimes can affect smart home users. Extortion campaigns in the form of ransomware are abundant today, and there is nothing indicating that it will be less frequent in the future. This category of crime includes ransomware, where critical parts of the computer or IoT system are locked or encrypted, and a ransom is demanded to give access to the locked parts. Ransomware typically targets documents, photos, and other file types containing valuable information, but in a smart house environment, it may target climate control systems or other systems that are both critical, hard, or expensive to repair or change. An investigation of this type of case will be the same as the computer ransomware counterpart: Find the attack vector used, analyze malware for weaknesses in the encryption, attribute the code or infrastructure to known groups, investigate the technical infrastructure of the malware, and follow the money track.

Another type of crime is cyber invasion of privacy and cyberbullying. There are many ways to invade the privacy of someone, from harvesting private information by online advertisers to invasive kinds of surveillance such as an attacker recording events happening in the home, recording audio, video, or other data from a smart home. Hello Barbie and Cayla, are examples of dolls marketed toward kids that can record audio, send it to a cloud for processing and keep a conversation with the kid. The recordings from Hello Barbie can, according to the terms of service, be sent to other third parties to help improve the voice recognition system, and the Cayla doll was hacked to change the responses and speak expletive words (Michael & Hayes, 2016).

There is also the so-called hacktivism or cybersecurity vigilantism, or crimes where the reward is based on an expressed ideology rather than the more traditional rewards, such as financial gain or social status. One example of this is the malware BrickerBot, that were bricking[3] IoT devices that were not properly secured. The reason for creating this malware was, according to the claimed author, to reduce the spread of other botnets by removing the vulnerable devices from the Internet before they could be infected (Goodin, 2017).

IoT promises an ecosystem of various systems and infrastructures. Some of these are low-cost devices that are deployed and will not be deprovisioned or retired, even after their intended usage is finished. For the well-managed systems, there exists a plan for deprovisioning that considers how it should be retired (Soos, 2018). For many systems, this is not the case, and old systems might still be connected to the networks without updates, risk assessments, or even without anyone knowing or remembering them. This means that there can be several attack vectors readily available for criminals to use.

At last, as environmental monitoring is gaining interest, these devices might help in detecting and investigating environmental crimes. The questions when it comes to the IoT systems can be of the precision and accuracy of the monitoring equipment, the integrity of the recorded data, whether the sensors have been moved, and their detection abilities. Some sensors can sample at specific times or intervals, or the sensors might post statistical data at specific intervals.

6.3.2 Identifying an IoT Investigation

When starting a cyber investigation, we might not know anything about the type of devices we will see in the case. As investigators, we need to be aware of our biases when it comes to what we are looking for. If we think we are looking for a PC, it is easy to overlook a small temperature sensor, even though it has the processing power and connectivity that we are looking for. One way of overcoming this bias is to think of the individual capacities we are looking for and then match them to the devices we find instead of looking for a particular type of device.

An example is we are looking for a webserver serving a set of simple, static pages. The needed capacities are that it has some way of a two-way connection and enough processing power and memory to get HTTP requests and deliver static pages out. This is not a particularly computationally intensive task, and can be served by, e.g., devices like a NodeMCU ESP8266 development board or a VoCore OpenWRT device, as shown in Figure 6.8.

At the beginning of an investigation, we do not necessarily know which direction the case will take and which technology we will be facing later. As we saw in Chapter 2, to be able to properly investigate all aspects of an IoT system, we can divide the investigation into several parts that need different skills and competence (Hunton, 2011b). The *modeling phase* of the investigation needs to keep the possible implications of the IoT system in mind, and the *assessment phase* will have to properly assess the available information about IoT systems.

During the *impact and risk phase,* the investigators should be cognizant of the unique characteristics of the particular IoT systems, and the *planning phase* needs to identify which new investigation steps should be taken, where data is stored, which expertise is needed, etc. The *tools phase* is selected based on the plan, and new tools for the IoT system

---

3 Render the device unbootable, i.e., as functional as an expensive brick.

Figure 6.8 Small devices that can act as simple web servers or offer other services, here shown by a NodeMCU ESP8266 development board and a VoCore OpenWRT device. (Photo copyright Jens-Petter Sandvik.)

might have to be developed or assembled. During the *action phase*, it is important that all involved know what to look for and which pitfalls to avoid. At last, we have the *outcome phase*, where the information is collated, evaluated, and presented. This model, as defined by Hunton, encompasses the digital forensics process and focuses on the investigative point-of-view rather than the more technological-focused digital forensic process.

For an investigation, it will be important to identify the IoT system and what type of information it might hold. The identification of the system might come from the initial information that starts the investigation, as the result of a testimony by a party, devices found at a particular site, or found by digital forensic analysis of other systems. The act of prioritizing which part of the system to collect evidence from and which traces to collect from a particular device or location is called *triage*. Triage also includes the selection of a proper acquisition method and assesses the probability of collecting evidence that can support or falsify both current and future investigation hypotheses.

Definition 6.6: Triage

Identify and collect the most relevant data as quickly as possible, i.e., identifying and prioritizing the most relevant evidence (Flaglien, 2018).

6.4   IoT Forensics

In an investigation involving IoT, we will naturally need data from these devices or systems. This is the digital forensic part of the investigation. In this section, we will first look at existing digital forensic areas and how they are used in IoT forensics, then describe the models created for IoT forensics, and in the end, describe new challenges that IoT introduces for the forensic investigation in these systems.

6.4.1 IoT and Existing Forensic Areas

The field of digital forensics contains many subfields, some of which overlap. As IoT itself covers many types of technologies, it is not surprising that such a forensic investigation also includes many of the digital forensic fields. Some of the fields described here are more thoroughly described in the book "Digital Forensics" (Årnes, 2018).

As many of the devices can be considered *embedded systems* and *mobile phones*, the forensic collection and examination of data from these devices will typically fall under the categories of *mobile and embedded systems forensic*. The challenge that is typically encountered in this field is the limited options for acquiring the raw data from the device, as many of the forensic tools do not handle uncommon operating systems or storage formats that are used in embedded devices. Some systems that are not well supported by forensic tools are operating systems like VxWorks, QNX, Contiki OS, or FreeRTOS.

One of the key elements of IoT is the networking and the Internet connectivity. This means that traces can be located in network logs, network flows, Intrusion Detection Systems (IDS), and other types of network equipment. *Network forensics* is a field within digital forensics that focuses on these types of traces. In addition, there will be evidence and traces to collect from the Internet, and *Internet forensics* is a part of this. For a deeper understanding of this field, please see, e.g., (Bjelland, 2018).

Data collected and processed by IoT systems are often stored or processed by *cloud* systems, which leads to another forensic specialization, namely *cloud forensics* (Ruan *et al.*, 2011). This is a branch that has a two-fold approach: on the one hand, it collects and examines data from virtualization platforms and cloud servers from the service platform owners' side in order to detect, e.g., deleted data from the service side. On the other hand, it also can collect data from the provided APIs from the Internet side of the service as a user. Figure 6.9 shows an example of data in a Google account, where the data has come from a voice command to a Google home device.

Of course, an IoT system consists of computers in addition to the things themselves, and *computer forensics* is still an important part of IoT forensics (Hamm, 2018). Computer forensics was earlier used as a synonym for digital forensics but is now an umbrella term for the forensic processes that focus on data from an ordinary computer. Computer forensics can split further into fields such as *file system forensics* and *database forensics*, among others, two fields that are similar, where the two fields are separated by the type of objects being stored. A file system and a database are in many ways similar: They both store data objects such that they can be searched, written, and read efficiently. The optimizations and usage of the various formats can be very different between them.

There is also the concept of *live forensics* that is also an important part of IoT forensics. Live forensics is the acquisition and examination of running systems and covers the acquisition and examination of the RAM directly and to run diagnostics tools, monitoring tools, triage tools, or other applications for collecting data from the system. The challenge in performing a live acquisition is that the tasks will render a footprint of the activity on the system and risk overwriting data while it is collecting traces. We must carefully consider and understand the impacts of our actions on the system and assess

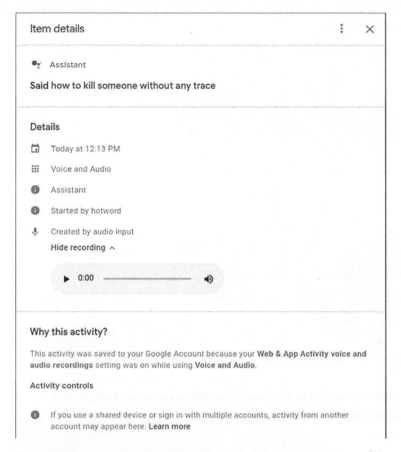

Figure 6.9 Data processed in the cloud from a Google Home speaker.

which parts of the system that can be changed in order to perform a live acquisition. This means carefully assessing the *order of volatility* in the system.

We also must include *computational forensics*. This is the use of computational methods for forensics, and it is not necessarily tied to digital forensics but can be used for all types of forensic sciences. This field can use simulations, statistics, and machine learning methods in order to analyze the traces found during an investigation (Franke & Srihari, 2008).

To summarize, we can look at the simplified relationship between the fields in Figure 6.10. Should the term IoT forensics cover everything connected with IoT, or should it be used only for the forensic actions we take that are not covered in the other fields? There is no exact answer to this, but it might be useful to keep in mind that most of the IoT technology in use today is well-known technology. An investigator should not think that in order to perform an IoT investigation or forensics, she must throw out all existing knowledge and learn everything from scratch, but rather think of it as a small incremental addition to the existing knowledge. Just as IoT systems consist of already existing technology set in a new context, forensic examinations of IoT systems also consist of established forensic methods. The importance of a forensically sound approach

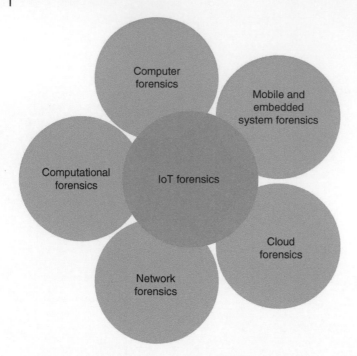

Figure 6.10 IoT forensics has a huge overlap with other established digital forensic fields.

for IoT systems is important such that *evidence integrity* and *chain of custody* are maintained for the collected evidence.

6.4.2 Models

The majority of IoT forensics research can be categorized into at least three main research areas:

- Frameworks for forensic processes
- Forensic readiness
- Forensic examinations of specific devices or systems

The first is frameworks for the forensic process (especially identification, collection, examination, and analysis). The second research area is frameworks for forensic readiness, or how IoT systems can store data that will be important for a forensic investigation after an incident is detected or an investigation that leads to an IoT system. The last research area is the more specific assessments and solutions for more concrete topics like specific acquisition methods for a particular set of devices or reverse engineering of file formats.

6.4.2.1 Frameworks for Forensic Processes

Two of the models that have been described for IoT forensic frameworks are *1-2-3 zones of Digital Forensics* and *Next Best Thing Model* (Oriwoh *et al.*, 2013). The term 1-2-3 zones refer to the locations where traces can be found and is meant as a tool to identify

Table 6.4 1-2-3 Zones of digital forensics.

Zone	Name	Description
Zone 1	Internal network	Devices, computers, etc., that are locally connected to the internal network
Zone 2	Middle	Gateways, routers, IDS/IPS, etc. The devices that operate between the internal and external network
Zone 3	External network	Cloud, web services, ISP data, etc. All locations where data from the internal network and the usage is stored

the traces. The zones are defined to be a categorization of the network, where each zone can be collected in parallel. Table 6.4 is an overview of the defined zones. Zone 1 is the internal zone of the network, where the devices themselves are found. Zone 2 is the parts of the network connecting the internal and external parts, which means gateways, routers, IDS logs, etc. Zone 3 is the external part of the network and spans cloud services, Internet service providers, and other service providers. The idea is that data from these zones can be collected in parallel or prioritized depending on the resources and information available in the case.

Definition 6.7: 1-2-3 Zones

A model where the traces can be collected from the devices themselves (zone 1), the gateways and routers (zone 2), and cloud services (zone 3).

The next-best-thing model is a triage model to find alternative traces in the system (Oriwoh *et al.*, 2013). If we know that a certain device might contain some evidence and we are not able to directly collect data from that device, we can instead look for evidence in devices or systems connected to the device of interest. An example of this can be a smoke detector; if we are not able to acquire data from it directly, we can look for the system where the smoke detector is connected, such as a home safety control system or a service provider for this system. If that data also is unavailable, we can look for the evidence in the next connected system, such as a mobile phone or other devices that are connected to the service provider.

Definition 6.8: Next-best-thing model

A model to describe the potential for traces to be found in devices connected to the device of interest.

6.4.2.2 Forensic Readiness

There have been a few models proposed for forensic readiness. These models assume that the system is connected in one network, typically a home network and that the owner or administrator wants to ensure forensic readiness in the system. In 2013, the Forensic Edge Management System (FEMS) was proposed by Oriwoh and Sant (2013). This is a system that targets a smart home environment. The system is based on a three-layer architecture:

1) Perception layer
2) Network layer
3) Application layer

The perception layer collects and analyzes data, the network layer is responsible for communication between the perception layer and the application layer, and the application layer is responsible for the user interface and user communication.

In addition to these layers, the system defines both security services and forensic services. The security services run continuously, monitoring the network for incidents, while the forensic services will start when the system detects an incident. The security services consist of the following services:

- Network monitoring
- Intrusion detection and prevention
- Data logging
- Threshold establishment

Network monitoring is defined in this framework as the monitoring of the network status. IDS and IPS systems will monitor the network traffic and log data. Threshold establishment is the part of the system that decides the thresholds for triggering the system's forensic services. This is a learning system that will either learn by itself based on the usage of the home automation system or set by the preferences of the user.

The forensic services are started when an incident is detected or suspected, and the system enters a trigger zone. The following services are defined for forensics:

- Data compression, parsing, and differentiation
- Timeline creation
- Alerting
- Preparation and presentation of results
- Storage

The data compression, parsing, and differentiation services select the relevant data showing the events based on the type of incident. The timeline creation will then create a temporal view of the events related to the incident. The alerting service will decide whether the incident is important enough for alerting the user. The final stage is to collect the stored information, prepare it and present the results of the events and actions taken by FEMS. The storage service is responsible for keeping secure storage for the forensic data such that the integrity is ensured.

In 2015, another system for forensic readiness was proposed. Forensic-Aware IoT (FAIoT) is a "centralized, trusted evidence repository" and consists of three modules (Zawoad & Hasan, 2015). These modules are:

- Secure evidence preservation module
- Secure provenance module
- API for evidence access

The secure evidence preservation module monitors the IoT devices and stores sensor data, network logs, network statistics, etc. As this can generate a huge amount of data, the authors propose to use Hadoop Distributed File System (HDFS) for this storage. In order to preserve confidentiality, the stored data should be encrypted.

The secure provenance module is responsible for the *chain of custody*. This module will log all requests to the data stored by the secure evidence preservation module. A provenance-aware file system is used for keeping a ledger of access and changes to the data in a provenance record (Muniswamy-Reddy *et al.*, 2006).

The investigator can access the data through an API that is only accessible by a few users. The original paper suggests that only investigators and the court should have access to this API.

Another framework proposed by Meffert *et al.* is the Forensic State Acquisition from Internet of Things (FSAIoT) (Meffert *et al.*, 2017). The framework consists of a Forensic State Acquisition Controller (FSAC) and state collection methods. The FSAC is implemented using the open-source OpenHAB[4] platform and will store the changes that the IoT devices report (i.e. the state of the things.) The data is only read from the devices and not written, timestamps are logged with the date, and the state report is hashed to ensure a forensically sound acquisition of the data.

Meffert *et al.* specify three methods for collecting the state data from devices. The first method is directly from the devices themselves and exemplifies this with an IP camera on the local network, where the FSAC can retrieve the state change directly from the device. The second method is to collect data from the cloud, as some devices only communicate with a cloud service, and the state information can be collected from this cloud service. The last method is to collect state information from another IoT controller that controls one or more IoT devices.

6.4.2.3 Forensic Examination of Specific Devices and Systems

The third research area is the forensic examination of IoT devices and IoT systems. The research in this area is often focusing on specific devices, specific brands, or specific hardware.

In the smart home environment, Google Nest[5] is a set of devices, ranging from indoor and outdoor cameras and doorbells to smoke detectors and thermostats. Dorai *et al.* (2018) examined a network with several Nest devices together with a Google Home speaker and the iOS companion app. Even though they did acquire data directly from the Nest devices, they were also able to analyze the events in the system based on the iOS companion app. This is a good example of the *next-best-thing model* as described earlier.

Chung *et al.* (2017) analyzed the Amazon Alexa[6] ecosystem. They analyzed both the cloud data from Amazon and data from the companion app. The cloud data was acquired by using unofficial APIs, and the companion app data was acquired from a mobile phone. As the companion app is a web-based application, traces could also be found in the web cache of the device.[7]

---

4 Open Home Automation Bus: https://www.openhab.org
5 https://nest.com
6 https://www.amazon.com/Amazon-Echo-And-Alexa-Devices/b?ie=UTF8&node=9818047011
7 The WebView class is used for web-based application in Android.

Azhar and Bate used a three-stage examination for a smart home environment (Azhar & Bate, 2019). They used If This Then That (IFTTT) to set up interactions between the Google Nest, Phillips Hue, and Amazon Alexa, such that one device would trigger other events. They focused on the traces that they were able to collect with network dumps and from the cloud and not from the devices themselves. They found that there is a huge amount of data at the cloud services, and Hue made the information about all users of the hub accessible to every user.

Smart personal devices can be an important source of evidence as well, and Odom *et al.* performed a forensic examination of smartwatches (Odom *et al.*, 2019). They experimented on the two connection modes of the watches, where they could operate either in *connected mode* or *standalone mode*. The first mode is when the watch is connected to the paired mobile phone via Bluetooth or Wi-Fi, and the second is when it is connected to the cellular network, respectively. They found that much of the data could be found both in the watch and in the companion app, but a few differences existed, where some information was only on the originating device. This difference was also changing with the connection state.

Another location of evidence is in vehicles, and Le-Khac *et al.* examined electronic evidence in smart vehicles (Le-Khac *et al.*, 2018). They first examined the entertainment system from the on-board diagnostics (OBD) port and by a chip-off method of the nonvolatile memory of the multimedia system. While the OBD port gave diagnostics information and technical information about the car, the multimedia system could reveal at least the last GPS fix of the vehicle. Other types of information that can be found in the multimedia system are contact lists, call logs, and other data shared with connected mobile phones, typed in addresses, etc.

Other researchers have focused on the radio chips in order to extract information from the nonvolatile memory of a Z-Wave ZW0301 chip (Badenhop *et al.*, 2016). The researchers could read out the content of the flash memory using an exposed Serial Peripheral Interface (SPI) interface on the package. They could establish pairings that had been performed and information about nodes in the vicinity. For an investigation that focuses on the network part of suspected malware attacks, this is often relevant information.

As mentioned in Section 6.2.4.3, the MQTT protocol is often used in IoT systems, and Shah, Rajdev, and Kotak did a forensic analysis of various clients and brokers (Shah *et al.*, 2019). They performed a network analysis of the traffic, examined network dumps, and found that each implementation left its own set of traces. For example, topic names and messages were found in all examined brokers but only found in two of the three examined clients.

These are just a few examples of the papers that have been examining IoT systems and devices. As an investigator, it can be very useful when faced with an unknown system to do a search and see if other researchers have examined the system before.

6.4.3 New Forensic Challenges

Looking at each individual part of an IoT system, we see that many of the elements have existing and well-established methods for forensic investigations, as described in Figure 6.10. A system might consist of some sensors and actuators, which can be regarded as an embedded system, a router, a cloud service, and a mobile phone. Each of these contains some type of storage that can be forensically acquired, and they will communicate, generating network traffic. There are, however, both new challenges and existing challenges that are augmented by new IoT technology.

There are also additional considerations that need to be assessed before starting the forensic process. The size of the objects can be small, both when it comes to physical size but also memory size. The location of the devices can also be a challenge because they can be in inaccessible locations or are not necessarily in the vicinity of the crime scene. The relevance of the evidence on a particular device can be hard to assess before we have examined the data. Legal issues might arise, especially when it comes to jurisdiction. Network boundaries can be blurry, so there is not necessarily a clearly defined perimeter. At last, it is the question about tools, whether they are able to collect and examine data, but also whether the device or system is hardened and has implemented physical and/or digital security (MacDermott *et al.*, 2018). The smaller memory sizes and the greater number of peripheral devices mean that it will be more time and resource demanding to collect a smaller amount of data.

6.4.3.1 Identification

As many IoT systems are designed to be ubiquitous, wireless, and inconspicuous, they might be hard to identify. One way is to look at the network traffic, the DHCP logs, ARP logs, or other network logs in order to map the network and find devices. It should be noted that devices do not necessarily show up in the network logs, as they might be connected to other networks and communicate over, e.g., a cloud infrastructure or be on the other side of a gateway, where the gateway communicates with the IP infrastructure. An example of this is a Bluetooth-connected device that uses another embedded system as a gateway between the device and the Internet side.

For some environments, just scanning various logs for devices might not be enough, and instead, we can scan the radio frequency (RF) spectrum for devices. Bouchaud, Grimaud, and Vantroys proposed to passively scan for devices that are actively sending, and if the scanning occurs for a long enough period, it can also pick up the signals of devices that only periodically transmit anything. For passive devices or devices that are passive for a longer period than the time we are scanning, the authors propose to generate traffic by, e.g., service discovery protocols or generate network changes by adding, jamming, or isolating devices. The idea is that devices that are passively listening will generate new traffic that can be analyzed. The authors also propose to use the physical RF signal *signatures* in addition to their protocol signatures in order to recognize the type of device in the network (Bouchaud *et al.*, 2018).

6.4.3.2 Volume, Variety, Volatility

Three existing challenges that will increase in the future are the quantity of data, the heterogeneity, and the lifetime of evidence data. As more devices are used, they generate more data, and various types of devices will produce, process, and store this in various formats. These three challenges are named Volume, Variety, and Volatility, and can be recognized as the three original V's of big data.[8] Thus, the amount of generated data from IoT systems can be considered big data. Each device might not store much data,

---

8 The three V's of big data, soon became 4,5,6,7,8, 10 and up to 42 V's. The latter set of V's was a joke posted at a blog, pointing at the trend to find new words starting with V for describing big data.

so collecting data from the individual devices will be time-consuming, and very little data can be collected from each device. So, we have data from the peripheral system that will be more resource demanding to collect, and heterogeneous systems will also create challenges to collect data from a variety of technologies. The centrally stored data will both be of a huge volume and in a variety of formats. So, on the one hand, we have more devices that are resource demanding to forensically extract information from. We also have more data and a wide variety of data formats that must be interpreted.

The lifetime of the data in an IoT system can be much shorter than in other systems. We say *can* here because it will not necessarily be a shorter lifetime, as it depends on the implementation and which type of data that are considered, and the size and type of storage. Raw sensor data will probably be very volatile, as it is processed and the raw data is overwritten by new measurements, but processed data might live for a long time. The processing might also include anonymization of the data and extracted statistics, such that it will be less useful for an investigation.

The analysis phase of IoT forensics often consists of a huge amount of information that should be analyzed in order to correlate events and find evidence about the events. Data reduction can be an important step to reduce the time and resources used for getting a quick overview of a case. One method for such a data reduction scheme can be to instead of using the raw image files, preprocess the files and then extract information about files systems, a predefined set of files, pictures, documents, etc. After this preprocessing, programs like the Bulk Extractor can be used to extract certain information before storing the results in a logical evidence file (Quick & Choo, 2016, 2018). Quick and Choo reported that the time used for opening 2.7 TiB[9] of evidence files from 18 computers and other devices with the forensic tool NUIX decreased from 107 hours for opening the raw files to 13 hours for opening the reduced files. The preprocessing of the data will still take time, though.

6.4.3.3 Fog Computing

Another new challenge that will come is fog computing. This is a term used for describing local clouds, where data can move between these local clouds, or fogs, as devices move between networks. The reason for introducing local clouds is that much of the processing will happen closer to the edges of the networks, and a platform to virtualize the local storage solutions allows for both lower latency to the data and movement of data between fogs (Bonomi *et al.*, 2012). Some of the application areas for fog computing described in the paper are:

- Smart grids
- Connected vehicles
- WSNs
- Smart building control
- IoT and CPSs
- Software-Defined Networks (SDNs)

---

9 Tebibytes, 2^{40} bytes of data, as opposed to a terabyte, which is 10^{12} bytes of data (IEEE, 2003).

Forensic challenges in fog computing environments are in many regards the same as cloud computing challenges: the dependency on cloud service providers, integrity preservation, forensic tool limitations, logs to examine, chain of custody, and legal issues, especially when it comes to jurisdiction (Wang *et al.*, 2015). The more specific challenges for fog forensics rise because of the low bandwidth, highly localized locations of the evidence, and more logs to examine and in more locations. Evidence dynamics is also an issue with fog data, as data can be changed during its lifetime.

6.4.3.4 Unreliable Data

As many low-cost devices get introduced and start generating or processing data, the risk of failures will increase. Some of the failures will render the device in a condition where it does not work at all, but some might fail undetected. It will still feed data into the system, but the data itself might be unreliable. A forensic investigation will not necessarily detect this unless the data clearly shows absurd results or does n'ot fit the investigation hypotheses. Casey showed the importance of describing the uncertainties to examinations and proposed to use a seven-segmented scale to describe the certainty of the digital evidence (Casey, 2002). This scale is reproduced in Table 6.5.

Data can be unreliable for various reasons, too many to enumerate and describe here, but some generalizations can be made for the sake of understanding. In dependability analysis, there is a concept of fault-error-failure-chains. This means that latent *faults* in the system activate or lead to *errors* which again cause or propagate to *failures*. Errors are where the system is in an unintended state, but it still provides the requested service. The errors can lead to failures where the services the system provides do not work as intended, and the system does not provide the requested service. A failure in a system can trigger other faults, which again lead to errors and failures.

This model is a good representation to start analyzing the correctness of our data. The failure in our case is the wrong data being presented to us and can, for example, be categorized in *deterministic and nondeterministic failures*. A deterministic failure gives the same result for the same inputs to the system. Nondeterministic failures depend on both the specific errors and their timing, which can make the failure seem to be one of a kind or hard to reproduce (Trivedi *et al.*, 2010). There are also categorizations for the

Table 6.5 Levels of certainty related to evidence, as proposed by Casey (2002).

Level	Description	Usage
C0	Contradictory evidence	Erroneous/incorrect
C1	Very questionable evidence	Highly uncertain
C2	Single source evidence, not tamper-proof	Somewhat uncertain
C3	Unexplained inconsistencies from tamper-resistant sources	Possible
C4	Single source, tamper-proof source of evidence or from multiple, not tamper-proof sources	Probable
C5	Multiple, independent, tamper-proof sources of evidence, with small inconsistencies or lacking data	Almost certain
C6	Unquestionable, tamper-proof evidence	Certain

type of faults causing the errors that trigger the failure. One classification from Avizienis *et al.* is shown in Table 6.6 (Avizienis *et al.*, 2004), and can be used by an investigator to establish hypotheses about the causes of the failures and help cover the possible hypothesis space.

One example of unreliable data is where clocks are seen to jump in time if the battery voltage falls under a threshold for a certain time in some CPUs (Sandvik & Årnes, 2018). This behavior was detected in a criminal case and triggered research into the topic, and several devices were tested. One of the tested Android phones would adjust the clock up to 12 years into the future when the voltage was held at approximately 2.0–2.1 V for 10 seconds. The authors did not go into details of the cause for this behavior, but a suspicion is that the registers in the processor holding the current time will start to experience bit errors when the voltage is too low to sustain the state of the register. In the end, the registers were reset, and the clock reset to the epoch of the operating system.

Other vague and seemingly random failures might arise from devices that have been exposed to electrostatic discharges, moisture, temperatures outside the operating window, radiation, mechanical damage to wires or soldering joints, or software bugs triggered by an unusual combination of inputs.

In addition to unreliable data stemming from failing systems, we have another source of inaccuracies in the data, namely imprecise and inaccurate data from working systems. We must define these terms, as they are often used interchangeably but technically have different meanings. The *precision* describes how much spread there is to the measurements. The accuracy, on the other hand, describes how far the measurement is

Table 6.6 Fault categorization (based on Avizienis *et al.* (2004)).

Fault class	Description	Fault types
Phase of creation or occurrence	When is the fault introduced?	Development
		Operational
System boundaries	Where is the origin of the fault?	Internal
		External
Phenomenological cause	What is the cause of the fault?	Natural
		Human-made
Dimension	Is it a software or a hardware fault?	Hardware
		Software
Objective	Is the fault introduced with a malicious objective?	Malicious
		Nonmalicious
Intent	Is the fault planned or nonintentional?	Deliberate
		Nondeliberate
Capability	Did the fault get introduced as a result of (human) lack of competence and/or organizational shortcomings, or was it accidental?	Accidental
		Incompetence
Persistence	Is the fault present permanently or just for a period of time?	Permanent
		Transient

from the actual value. So, we can have good precision but low accuracy, and be consistently wrong, or a high accuracy but low precision and be averagely right with low certainty.

Figure 6.11 shows this relationship. As IoT systems often can sense their analog environment and digitize this environment to digital values, the precision, and accuracy of this measurement can be off. A temperature sensor for a smart home might show 19 °C, while the temperature, in fact, is 23 °C. For a homeowner, this might not mean anything more than that he must adjust the temperature by a few degrees, so it feels comfortable, but if this value is used for, e.g., calculating the time of death of a body found in the apartment, the whole investigation might rely on the wrong timeline of events.

Please note that the cause of the accuracy and precision is often referred to as systematic and random errors. Systematic errors are errors that are caused by a common fault, e.g., an uncalibrated thermometer that always shows 1 degree less than the actual temperature or a broken assumption about the model used for describing the readings. Random errors, on the other hand, are, e.g., the variability caused by other, seemingly random causes. This might be random fluctuations of colder and warmer air affecting the thermometer or quantization errors introduced by categorizing continuous data in discrete categories.

6.4.3.5 Dynamic Systems

IoT systems can be highly dynamic, where changes happen continuously. The changes can be to the configuration of the system, the network structure, the location of the cloud or fog data, devices can move between networks, devices can lose power or be set in power-saving modes, etc. This means that in order to interpret the evidence in the system, we might have to recreate the state of the system or

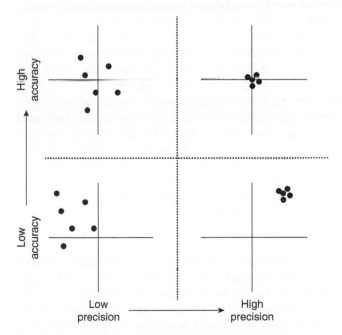

Figure 6.11 The relation between precision and accuracy.

parts of the system before the evidence gives a truthful picture of the historical events or the investigated incident.

How can the investigation recreate the state of a system that has changed and probably is changing as the identification and acquisition are being performed? First and foremost, we must be aware of which systems are dynamic. Then we must assess the impact it has on the investigation hypotheses. A thing can be removed from or added to a network, and there might not be any logs of the network changes. A VANET does not necessarily log which vehicles have routed packets for a vehicle of interest, but how the *exact* routing was at the time is not necessarily relevant for the investigation, as long as the network traffic *could* have been routed between the two endpoints.

6.4.3.6 Data Fragmentation

One of the challenges that will increase is data fragmentation. In order to get the full dataset from an event, we must find the evidence in several locations. Some might be in the devices, some in the gateways, some in the companion app of a mobile phone, and some in the cloud. The locations of the data have different complexity for collecting evidence and might be above the level that a forensic lab is able to collect or examine.

If the data is stored in cloudlets, the data might be spread over several of the cloudlets, and in order to get all data, we must find all cloudlets that the devices have communicated with. Another fragmentation is in virtual systems, typically cloud solutions, where traces of the virtual environment can be spread over several physical machines.

Sometimes we do not need all possible data, but sometimes we do not know what we are lacking and might end up with wrong conclusions.

6.4.3.7 Multiple Technologies and Standards

When IoT started gaining traction, many thought that the technologies and emerging standards would converge towards an "IoT standard." This was a view that do not seem to have become a reality. The challenges addressed by the various application areas are different, and different solutions are needed. A home automation system might be connected to the Wi-Fi and power outlets in the home, while an environmental monitoring system needs long-range data connections that can run on batteries for a long time. Agricultural applications might have other needs, such as precise positional systems for robots, where the sensor data from, e.g., moisture sensors do not need to be in real-time.

As there is no single "IoT technology," there is not a single "IoT investigation" method that covers all aspects of investigations of IoT systems. We rather must look carefully at the information in the investigation, the hypotheses proposed, and the system at hand. With many types of cybercrime, the symptoms that one party sees can be just a part of a bigger scheme. A Telnet login attack against a nonsensitive device can be a part of creating an infrastructure for enabling other crimes.

6.4.3.8 Jurisdiction

As cloud computing is an integral part of the IoT paradigm, the same challenges that we see in cloud forensics are also seen in IoT forensics. One of these challenges is the location of the evidence, which is often found in locations governed by other jurisdictions. The Google Home loudspeaker does not store much information in the device, but most is sent to Google's cloud system and stored there, as the example in Figure 6.9 shows.

Sometimes the user can access all his or her data, but often the user only has a limited set of the data. To get the rest of the data, a request must be sent to the cloud provider. Sometimes the user can make a request for the information, e.g., for cloud providers that follow the General Data Protection Regulation (GDPR) (European Council, 2016). The police or law enforcement agencies can request information from cloud providers. The request can be done through international cooperation channels created for this, or through the courts and ministries of the involved countries. The latter is a long process that, in the worst case, can take many months before the request is answered. More of the laws governing international investigation can be found in Chapter 3.

6.5 Summary

In this chapter, we introduced the concept of IoT, a short historical note, and some of the technological descriptions of what this is. There are many application areas for IoT, each with their own set of technology challenges. If we do not understand what IoT is and how it works, it is hard to find and interpret the evidence correctly.

We saw how IPv6 and the IETF IoT protocols were important for communication in IoT systems and how extra protocols have been introduced for adapting IPv6 to IoT systems. 6LoWPAN is a fragmentation and reassembly adaption layer for IPv6, and CoAP is a binary protocol for transmitting data between applications.

There are many existing digital forensic fields that are used for investigating IoT systems. Some of these are mobile and embedded system forensics, network forensics, Internet forensics, cloud forensics, computer forensics, file system forensics, database forensics, live forensics, memory forensics, and computational forensics.

We have looked at some of the forensic models, both active forensics models as the 1-2-3 zones and next-best-thing model, and models for forensic readiness, such as FEMS, FAIoT, and FSAIoT. The new challenges for forensics are the amount and diversity of devices and data, the volatility of the data, Fog computing, unreliable data, and dynamic systems.

6.6 Exercises

1 You are called out for a suspected murder in an apartment in a new building block. What are the possible locations of evidence, which type of evidence can be in these locations, and how should you proceed to collect the evidence? What is the volatility of the data in the various evidence sources?

2 What are the assumptions you have used for the previous question, and how will a change of the assumptions affect the conclusion from the answer given?

3 What are examples of operating systems used in IoT systems that are not well supported by forensic tools?

4 What is the relation between accuracy and precision, and which impacts can these have on evidence?

5   You are summoned to investigate a traffic accident that just happened, where a car has been speeding through a red light and ended up in a ditch right after the cross-road. Arriving at the scene, the driver tells you that the car was speeding by itself, and he was unable to control the car, and it steered itself toward the ditch. How would you approach the investigation? What are the hypotheses? How should the forensic work on the car be performed?

6   If you have any cloud-connected IoT devices at home, find out how to download your own data. What can this data tell about the operating environment and the usage of the system?

7   Why can a network of low-power and resource-constrained nodes in many cases not use IPv6 addresses directly? What is one solution described in this chapter to this problem?

8   How can uncertainties in the investigation be reported? Are there any models for describing these uncertainties?

9   Discussion: Does the General Data Protection Regulation make it easier or more difficult to investigate IoT systems? Why?

10   What is the difference between CoAP and HTTP?

7

Multimedia Forensics

Jens-Petter Sandvik[1,2,]* *and Lasse Øverlier*[1,3,]**

[1] Department of Information Security and Communication Technology, Faculty of Information Technology and Electrical Engineering, Norwegian University of Science and Technology (NTNU), Trondheim, Norway
[2] Digital Forensics, National Cybercrime Center/NC3, National Criminal Investigation Service/Kripos, Oslo, Norway
[3] Norwegian Defence Research Establishment (FFI), Gjøvik, Norway

Multimedia forensics is an umbrella term covering a variety of methods, objectives, and content. Image forensics may be the most well-known topic within multimedia forensics, as images are found virtually everywhere and often subject to editing. Serial numbers, camera settings, and Global Positioning System (GPS) coordinates are often found in the metadata structures of the photo. They are often used in forensic examinations to answer questions about the location of scenes and the time the image was taken and processed. This application is, of course, one part of multimedia forensics, but the focus in this chapter is what the media content itself can reveal about its history.

Video forensics uses many of the same techniques as photo forensics but also uses the temporal dimension of the video. The compression in videos is higher, making some methods used in image forensics imprecise. A video typically has an audio track that can be forensically examined using audio forensics techniques.

Deepfakes is a generic term used in particular for visual content that has been generated or manipulated and is hard to discern from actual, captured content. This term has been popular in media lately, as the technology has become better, creating photorealistic content at a budget most can afford. As the computing power available for most people has rocketed, the possibility for everyone to artificially generate images that looks like real photographs has also increased. In the future, we might have to question the authenticity of photographic or videographic content, not only for content where resourceful adversaries are involved but also for content where anyone has something to gain by manipulating the content.

Audio and signal forensics are other aspects of multimedia forensics. Audio forensics can be seen as a subset of signal forensics with one signal (mono), two signals (right and left channel), or more (surround, ambisonic), where the signal is an audio signal within the audible spectrum.

---

* Jens-Petter Sandvik is a Ph.D. student in Digital Forensics at NTNU and a Senior Engineer at the Norwegian Criminal Investigation Service (Kripos). Email: jens.p.sandvik@ntnu.no
** Lasse Øverlier is an Associate Professor at the Norwegian University of Science and Technology (NTNU) and a Principal Scientist at the Norwegian Defence Research Establishment (FFI). Email: lasse.overlier@ntnu.no

7.1   Metadata

While this chapter does not go into the details of metadata, it is still a topic that needs to be covered for a survey on multimedia forensics to be complete. Metadata can be data about the file container as seen from the operating system, but this we will consider within the computer forensics realm. This metadata is timestamps set when the file is created, accessed, and updated in the file system. File names and file system blocks containing the contents are also a part of this metadata. These metadata structures are artifacts found in the file system comprising the media file, and they will not necessarily follow the file as it is moved between file systems.

The second type of metadata for a multimedia file is the data attached to the container format of the file. A multimedia file usually has a container file format, such as Matroska or MP4, that contains the multimedia streams together with a description of their internal timings and the contents. Some file formats, such as JPG, have embedded the data streams and the container format in the same specification. In contrast, other file formats define the container format and allow several media stream types to be present. The metadata in the container formats can be present in many forms. For example, audio files, such as the MP3 format, might store information in ID3 tags. Various image formats store information about the image, such as the camera model, serial number, or GPS data in EXIF tags.

Some standard metadata formats are Exchangeable Image File Format (EXIF), International Press Telecommunications Council (IPTC) Photo Metadata, and Extensible Metadata Platform (XMP) for Joint Photographic Experts Group (JPEG) files, and ID3 for MP3 audio files. In addition, container-specific metadata structures are used in other file formats. Figure 7.1 shows the output from the tool *ExifTool* that prints out metadata for a wide variety of multimedia file types.

The metadata is generated by the software or firmware that processes the media file and is easily manipulated. Therefore, an investigator needs to carefully assess the risk of manipulated metadata for the data file under investigation. The metadata does not necessarily need to be intentionally manipulated. Still, it can be imprecise, such as the location in a photo, if the camera did not receive signals from the Global Navigation Satellite System (GNSS) service. Timestamps from GNSS satellites are often treated as precise. Still, the process of writing to a metadata timestamp can be imprecise and show a deviation of at least several seconds (Sandvik & Årnes, 2018).

The examination of metadata is an integral part of an investigation, and even though it is not the target here, an investigator should not forget it.

7.2   Image Forensics

Photo forensics is the term used for the forensic examination of photographs. This chapter uses the terms *photo*, *photography*, and *image* interchangeably. In digital forensics, the term image is also used for disk images and files containing other acquired memory devices. Even though the term is used to describe photographs and is used in many books and papers, it can be confusing from a digital forensic perspective to use it

[ExifTool]	ExifTool Version Number	:	12.05
[File]	File Name	:	20201128-1324-Hjorteparken-221.cr2
[File]	Directory	:	Pictures/2020/20201128
[File]	File Size	:	38 MB
[File]	File Modification Date/Time	:	2020:11:28 14:24:22+01:00
[File]	File Access Date/Time	:	2020:03:23 14:49:09+01:00
[File]	File Inode Change Data/Time	:	2020:11:28 19:15:04+01:00
[File]	File Permissions	:	rw-r--r--
[File]	File Type	:	CR2
[File]	File Type Extension	:	cr2
[File]	MIME Type	:	image/x-canon-cr2
[File]	Exif Byte order	:	Little-endian (Intel, II)
[EXIF]	Image Width	:	6720
[EXIF]	Image Height	:	4480
[EXIF]	Bits Per Sample	:	8 8 8
[EXIF]	Compression	:	JPEG (old-style)
[EXIF]	Make	:	Canon
[EXIF]	Camera Model Name	:	Canon EOS 5D Mark IV
[EXIF]	Preview Image Start	:	67928
[EXIF]	Orientation	:	Horizontal (normal)
[EXIF]	Preview Image Length	:	3052874
[EXIF]	X Resolution	:	72
[EXIF]	Y Resolution	:	72
[EXIF]	Resolution Unit	:	inches
[EXIF]	Modify Date	:	2020:11:28 13:24:22
[EXIF]	Serial Number	:	
[EXIF]	Lens Info	:	50mm f/0
[EXIF]	Lens Model	:	EF50mm f/1.8 STM
[EXIF]	Lens Serial Number	:	
[EXIF]	GPS Version ID	:	2.3.0.0
[EXIF]	GPS Latitude Ref	:	North
[EXIF]	GPS Longitude Ref	:	East
[EXIF]	GPS Altitude Ref	:	Above Sea Lelvel
[EXIF]	GPS Time Stamp	:	12:24:15.912
[EXIF]	GPS Satellites	:	6
[EXIF]	GPS Status	:	Measurement Active
[EXIF]	GPS Measure Mode	:	3-Dimensional Measurement
[EXIF]	GPS Dilution of Precision	:	2.2
[EXIF]	GPS Map Datum	:	WGS-84
[EXIF]	GPS Date Stamp	:	2020:11:28
[EXIF]	Thumbnail Offset	:	54792
[EXIF]	Thumbnail Length	:	13131

Figure 7.1 Some of the file and EXIF metadata from a photo taken by a DSLR camera, as interpreted by ExifTool.

for both graphical media content and the result of a memory device acquisition. The term *photograph* and its shorthand *photo* do not cover all visual media types. This is a term used to acquire images from an illuminated scene by an optical sensor system, either to a photosensitive substance or a photosensitive electronic sensor, excluding synthetic images.

This subsection starts with the photographic process, explains which parts of the process leave traces in a photo, and then goes into detail about examinations from each of the process steps.

7.2.1 Image Trustworthiness

An image seldom comes with a chain of custody and proof of the integrity of the image. Even though the history of an image is unknown, humans tend to place more trust in a picture than written or orally transmitted information.

However, an investigator can exploit information from the photographic process to help verify the source. By actively adding data to an image, the investigator can ensure the integrity of the original. A fragile watermark is difficult to alter without detection, making it hard to edit the image without leaving evident traces of manipulation.

Another way to ensure the integrity and offer non-repudiation is to add a cryptographic digital signature to the image. The signature can be traced to a known source, and any manipulation of the contents will render the signature invalid.

If the information to ensure the integrity is not actively added to the image at the source, the integrity of the image can still be examined. In addition, even though the information is not actively added, many of the stages in the image creation process are still passively adding fingerprints to the image. These can be analyzed to identify the process steps and parameters and find evidence of tampering.

7.2.2 Types of Examinations

Photos have been around for almost 200 years, and the concept of manipulating images was introduced not long after this[1]. However, today, with a vast number of cameras and photographers taking and sharing digital photos with available and affordable photo editing software, the correctness and authenticity of an image might be questioned.

Image forensics is a field within digital forensics that focuses on the content of photographs. In this chapter, the definition will be limited to photos taken with digital sensors and digitally stored. A photo can be examined to answer many types of questions. A classification of the forensic examinations is:

— Authenticity validation: Decide whether the photo is authentic or has been manipulated in any way.
— Source device identification: Identify the source device used for acquiring the photo.
— Photogrammetry: Measure parameters such as sizes, lengths, distances, and angles between elements in the photo, mapping it to the three-dimensional scene.
— Object identification: Identify objects in the photo.
— Spatiotemporal location identification: Identify the location of when and where the photo has been taken.

Establishing the authenticity of a photograph and identifying the source device are probably the most researched areas of image forensics. Photogrammetry is an area used for geographical surveys, and drone mapping is one of the applications of this. It is also an essential part of interpreting the contents of a picture to measure sizes, distances, and angles in the scene. Identifying objects in a photograph is a crucial task of many criminal investigations, especially in child abuse cases. Object and location identification both rely on photogrammetry and other information for identification.

The term *identification* is generically used in the list above to mean loosely "the determination whether an artifact belongs to a class, where the number of elements in the class

---

1 http://web.archive.org/web/20201101055916/https://www.cc.gatech.edu/~beki/cs4001/history.pdf, visited 2021-04-06.

can be from one and upward." This means that it is both to identify an artifact, location, or object uniquely and to classify the artifact, location, or object into categories or groups.

Another category not in this list because it has not had any known forensic application yet, is photometry. This is the measurement of light intensity from objects in photos and is typically used for measuring the light intensity from distant celestial objects in observational astronomy. For classifying the type of lamp used, a forensic application might be, for example, to find the level of illumination from a distant car headlight.

The technical, forensic examination of images can also be thought of as identifying fingerprints of the history of the image, just as traditional forensic examinations such as forensic ballistics both measures the traces and compares the striation marks to known marks from a bullet fired by a gun of interest in a controlled setting.

Instead of categorizing the forensic process in terms of the type of investigations, we can organize the process in terms of where in the photographic process the fingerprints are created. The three process steps that generate fingerprints are shown in Figure 7.2 by the colored boxes and are:

− Acquisition fingerprints
− Coding fingerprints
− Editing fingerprints

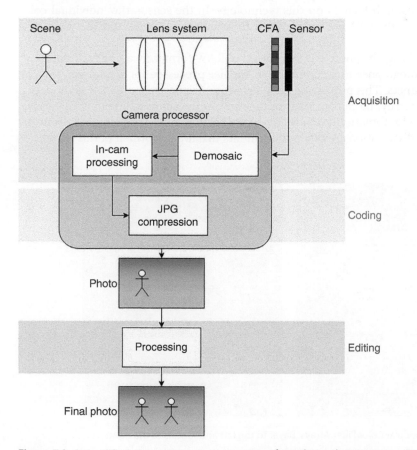

Figure 7.2 A simplified digital image capture process from the analog scene to a digital photo.

7.2.3 Photography Process Flow

Capturing an image will leave information about the process in the digital image. To understand what this information is and how to interpret and detect it, we need an in-depth understanding of the capture process and which types of traces are left by each of the process steps.

Figure 7.2 shows the capture process for a digital image. The photons reflected and refracted from a physical scene travel through a set of lenses. Each type of camera objective has its own set of lenses and materials used for the lens and the coating.

As the sensor typically is sensitive to a wide range of wavelengths, covering the whole visible spectrum, color filters are needed to distinguish between the colors and create a color photograph. This color filter is called a color filter array (CFA) and can have different patterns. For example, Figure 7.3 shows two popular patterns, a Bayer filter, and an X-trans filter. As can be seen from the figure, each color pixel consists of four sensors, each capturing one of the colors in the filter, which means two things: The number of raw pixels in the sensor chip is four times the number of pixels in the resulting image, and each color is slightly offset from the other colors.

As the sensors pick up the filtered colors from the CFA, the photons are converted to an electrical charge in a complementary metal–oxide–semiconductor (CMOS) or charge-coupled device (CCD) chip. Today's most common camera sensor is the CMOS type, so most research focuses on this technology. In the sensor, the individual color components from the CFA are interpolated to form a single pixel, and this process is called demosaicing.

After demosaicing, the photo is either stored in RAW format, awaiting further adjustments by a photographer in a computer, or further processed for a final image to be stored in the camera. This processing includes white-balance adjustments, as the actual colors, as detected by the camera, are not the same as the colors a human will perceive. The human mind automatically adjusts the white balance in the scenes we see. Contrast adjustments and saturation are other adjustments that are performed in this stage.

Figure 7.3 Two popular Color Filter Arrays: Bayer to the left and X-trans to the right.

After the raw photo has been adjusted, the image is often compressed and encoded in a JPEG format. The JPEG encoding is a lossy compression, where information from the picture is lost. The information that is lost is carefully selected so that the human observer should not notice any loss in quality.

The pixels in the raw image are first converted to the YCbCr color space, where Y is the luminosity value, while Cb and Cr are the chroma components. The chroma components are then spatially downsampled, as the human eye is not good at detecting color variances as much as luminosity variances. Each color component is then shifted from an unsigned range to a signed range: $[0,2^b-1] \rightarrow [-2^{b-1},2^{b-1}-1]$, where b is the number of bits for each pixel and compressed individually.

The converted, downsampled, and shifted image is then split into blocks of 8×8 pixels, and each block is then transformed into the frequency domain by a discrete cosine transform (DCT):

Equation 7.1: Discrete cosine transform (DCT)

$$G_{u,v} = \frac{1}{4}\alpha(u)\alpha(v) \times \sum_{x=0}^{7}\sum_{y=0}^{7} g_{x,y} \cos\left[\frac{(2x+1)u\pi}{16}\right]\cos\left[\frac{(2y+1)v\pi}{16}\right]$$

where $\alpha(u)$ is a normalizing factor that has the value $\frac{1}{\sqrt{2}}$ when $u=0$ and 1 otherwise, and $g_{x,y}$ is the pixel values at (x,y).

The quantization step uses the frequency domain blocks and divides the values by a value defined by the quantization matrix. The standard does not specify the quantization matrix for each quality setting, but a scaled version of the example matrix from the standard is often used. There is a different matrix for the luminosity and chroma components, which means the chroma components will have a higher compression rate. This division is rounded to the nearest whole number and stored. In addition to the down-sampling of the chroma components, the quantization step is responsible for the information loss in the compressed image.

The last step to the compression is serializing the blocks and compressing the resulting bitstream with a lossless entropy encoding. A variable-length Huffman encoding is often used for this step and ensures that the sparse 8×8 blocks are well compressed.

The resulting original JPEG image can be further processed or manipulated and resaved as a new photo, typically in JPEG format again. This editing can include light or color adjustments, image cropping, deleting, or inserting parts from other images.

A survey of image forensics by Piva splits the process above into three distinct classes (Piva, 2013). The *acquisition phase* is where the forensic methods focus on the process up to the photo has been demosaiced, and in-camera processing has taken place. The methods targeting the *coding phase* focus on the artifacts from the image encoding to a compressed image format. The *editing phase* is the phase after the photo has been encoded to a compressed image format, and other photo editing methods have been used on the picture, and some forensic methods target this type of processing.

In the survey by Piva, two forensic questions were the focus. The first question is whether the claimed camera was the source device used to capture the photo. The second question is about the photo's authenticity and whether the image depicts the original scene.

7.2.4 Acquisition Fingerprints

Acquisition fingerprints stem from the lenses, optical filters, CFA, the analog-to-digital conversion in the photographic sensor, and the in-camera processing before the photo is encoded and stored. These fingerprints can be used to identify the original equipment used for taking the photo or for analyzing the image's authenticity.

The methods for identifying the source device for capturing a photo target the acquisition and coding phases of the photographic process, as described in Section 7.2.3 and shown in Figure 7.2. These two phases are where the source device will affect the photography and create traces that can be linked to the specific camera, the type of capture device used, or the kind of software and settings used.

The authenticity of a photo means establishing that the photo is the original and has not been manipulated in any way to misrepresent it. Such manipulation can be, among others, copying parts of another image into the image, copying parts of the original photo and moving it to another location in the image, and stretching or shortening distances between points.

7.2.4.1 Lens Artifacts

The lenses in a camera are the first elements that affect the photons from a scene as they travel toward the sensor. As the light passes through a lens, it will be refracted, and depending on the angle the light reaches the lens, the light is refracted to a different angle through the lens. This effect is called aberration and will create unique patterns in the image that can identify at least the type of lens used.

The image field itself can be distorted by the lens or by the positioning of the objects. Barrel distortion is a type of distortion generated by the lens that causes lines in the image to curve around the center of the photo, and pincushion distortion is the opposite, where lines curve toward the corners of the image. A mustache distortion is a combination of these two.

Figure 7.4(A) shows a barrel distortion. Another geometrical aberration found in cameras is perspective distortion. This is caused by objects in the photo that take more or less space than in the natural scene. The seemingly impossibly big foreground in photos taken with fisheye or wide-angle lenses is caused by this and is referred to as an extension distortion. The opposite effect, called a compression distortion, is where the background seems very close and is present in photos taken with a telephoto lens.

Figure 7.4(B) shows chromatic aberration, where each color is refracted slightly differently through the glass. This occurs because each wavelength of light has a different refraction index and thus refracts with a different angle.

Aberrations can, just as in the source device detection, be used for detecting image manipulation. As the lens aberration follows the curvature of the lens, inconsistencies of the aberration in a photo can thus signify that a part of the photo has been copied in from somewhere else, possibly another photo.

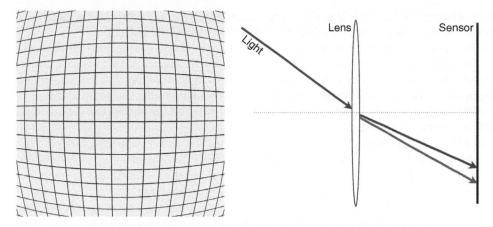

Figure 7.4 The distortions in a photo due to the effects of the lens (Figure by the author).

7.2.4.2 Color Filter Array

The CFA is another source of information that can identify a class of source devices. As described in Section 7.2.3, the CFA consists of an array of color filters in a pattern that will ensure that a sensor pixel only acquires photons with a specific wavelength, and a 2×2 sensor pixel array is then transformed to one color image pixel. The interpolation process to create one color image, also called demosaicing, leaves detectable traces on the image.

Some work on detecting the demosaicing effects aims at detecting the layout of the image filter and the demosaicing algorithm together with its parameters. Each camera brand and type has a specific set of such configurations, and the class can be identified by detecting this.

One of the methods used for detecting the artifacts from the CFA, described by Popescu and Farid, assumes a linear interpolation kernel and uses an expectation-maximization (EM) algorithm to estimate the parameters for the kernel (Popescu & Farid, 2005). The EM algorithm is a two-pass algorithm, where the first pass (the "E-step") calculates the probability for the pixel in the color channel to belong to a model, using Bayes' rule. The next step (the "M-step") estimates the form of the correlation between samples and the variance of the probability distribution from the previous step and updates a parameter of the prior step. The process iterates until the difference in parameter estimations reaches a given threshold.

The results show that CFA interpolation in images is detectable in the Fourier domain, where there is a strong periodicity for the signal. While the type of CFA algorithm can be harder to discern, the authors did see a particular grouping of the parameters. The method is less robust to JPEG compression than the methods using photoresponse non-uniformity (PRNU), but the false-positive rate is lower than the one using PRNU.

Another method is to first estimate the CFA pattern from the source device using the lowest mean square error among precalculated possible patterns or to use the fact that the directly acquired pixels will leave a different PRNU noise pattern than the interpolated ones. The reason for this is that the interpolation effectively introduces a low-pass filter to the channel, thereby suppressing the high-frequency PRNU noise.

Other research has also trained a support vector machine (SVM) to detect the camera model, where the authors used two different sets of artifacts: an EM algorithm for non-smooth parts of the photo, and the second-order derivative of the rows, and detecting the peaks in the Fourier domain (Bayram *et al.*, 2006). The results showed that the SVM trained on the combined set of artifacts could distinguish three camera models with 96% accuracy.

In addition to identifying the type of camera, the CFA can also be used for detecting manipulation of the image. Like inconsistencies in the PRNU identify regions where the original image has been manipulated, inconsistencies in the CFA artifacts in the photo also will show that the photo has been intentionally manipulated.

The method by Popescu and Farid, described earlier, can also be used for detecting areas that do not contain CFA interpolation artifacts or that contain artifacts not compatible with the CFA algorithm in the rest of the image (Popescu & Farid, 2005).

7.2.4.3 Sensor Artifacts

As the sensor is an analog-to-digital converter (ADC), where light intensities are encoded to binary data, the process is prone to noise sources in this conversion step. The most dominating source of such sensor pattern noise is the PRNU. This noise is a combination of several factors, where contributing factors are imperfections during manufacturing, flaws in the silicone, and thermal noise. The PRNU is usually stable over the chip's lifetime and can therefore be used to identify source devices for old photos or verify that two photos taken years apart come from the same device.

Chen *et al.* described in their paper a way of comparing the PRNU from a known device with the PRNU that is present in a photo under examination (Chen *et al.*, 2007). A model for the sensor noise is needed to compare the PRNU between a set of photos taken by a known device and the examined photo. This model simplifies the image signal and assumes that the signal in the photo consists of the actual signal, a zero-mean noise to the signal, and a combination of other noise sources. The signal and image pixels are here given as 1-dimensional vectors and not matrices:

Equation 7.2: Photo signal model

$$I = I_O + KI_O + \Psi$$

where **I** is the resulting photo signal, I_O is the incoming, noise-free light intensities, **K** is the zero-mean PRNU signal, and Ψ represents other random noise sources.[2]

The image data can be removed, such that we are left with only the noise components. This is done by subtracting a denoised version of the image:

---

2 Random in this context refers to the distribution of the noise from each source, not that the sources are randomly selected.

Equation 7.3: Image noise pattern

$$\mathbf{W} = \mathbf{I} - F(\mathbf{I}) = \mathbf{IK} + \Phi$$

where **W** is the resulting noise pattern, $F(\cdot)$ is the denoising filter, and Φ is the noise which consists of two other noise terms, namely the original Ψ and the noise introduced with the denoising filter, $F(\cdot)$.

Equation 7.4: Estimation of the PRNU signal, K

Given a set of N photos from the same camera, $\mathbf{I}_k, k \in N$, we can calculate $\mathbf{W}_k$, and from these two sets calculate the maximum likelihood predictor for **K**:

$$\mathbf{K} = \frac{\sum_{k=1}^{N} \mathbf{W}_k \mathbf{I}_k}{\sum_{k=1}^{N} (\mathbf{I}_k)^2}$$

This must be done with each device with which the examined photo will be compared. The source device can thus be identified by calculating the PRNU from Equation 7.3 for the examined photo and correlating the noise with the likelihood predictor for the devices in the reference set:

Equation 7.5: Image noise correlation

$$\rho_i = \mathbf{I}^0 \mathbf{K}_i \otimes \mathbf{W}^0$$

where the operator "$\otimes$" denotes a normalized correlation operator. The source device can then, according to the paper, be identified either as the one with the highest correlation or the one that is above a predefined threshold.

As the described method is the primary method of using PRNU for detecting the source device, there is much research ongoing to increase the method's robustness. The denoising filters are essential for removing the signal such that only the noise remains, as seen in Equation 7.3. A filter that is better at discarding the noise and leaving the signal will give a better noise signal when subtracted from the noisy image.

The PRNU can also detect the authenticity of a photo taken by a device. The idea behind this is that given a source device and a photo that should have been taken by this device, any deviations from the known PRNU will identify a non-authentic picture.

The deviation can be that a different source device has been used, that parts of the photo have been changed by copying from another photo, or that the parts have been copied from a different location in the same photo. The only editing that should be non-detectable is where two photos are taken with the same source device. Then the parts from the second image are copied into the first image at the exact pixel location, thereby preserving the PRNU.

Experiments by Chen *et al.* showed that a block-wise comparison of the PRNU in the photos showed that for 85% of the images, they could detect at least two-thirds of the manipulated regions, given a jpeg compression with a quality setting of 90 (Chen *et al.*, 2007). When the quality factor was set to 75, the number of correctly classified regions fell to at least two-thirds of the manipulated areas detected in about 73% of the photos. The false negatives were reportedly around the edges of the manipulated regions and in very dark spots in the photos, where the PRNU is suppressed. The false negatives were typically located in saturated areas with dark areas and a complex structure, where the signal is hard to discern from noise.

7.2.4.4 Other Image Acquisition Artifacts

Other artifacts from the image acquisition phase that imprints that leave traces in the photo have also been used for detecting the source device. One of these is the radiometric response of the sensor, which is characteristic of camera sensors of the same brand (Lin & Zhang, 2005). The radiometric response is the mapping of the scene's radiance to the intensity of the image pixels. This response is nonlinear to compress the dynamic range of the location to fit the dynamic range of the sensor.

The response function can be estimated and categorized by analyzing the histogram of the edges appearing in the image. Another method for identifying the response function is to use geometric invariants to get a consistent set of locally planar irradiance points (Ng *et al.*, 2007).

Also, the detection of re-acquisition of an image has been addressed, where an image is shown on a medium, either printed or on-screen, and a new photo of this has been taken. This poses new challenges as traces in the image that has been forged or manipulated are hidden by the recapturing process. However, artifacts like high-frequency specular noise indicate recapture that can be used to detect such re-acquisition.

Scanners also leave traces of the acquisition process in the image, and PRNU can be detected for these devices too, even though the way it acquires the image is slightly different, as it captures rows and combines the rows into a single photo. The contents of the scanned photos also tend to be documents rather than natural scenes, which introduces large, saturated areas to the scene, which makes the PRNU analysis harder.

7.2.5 Image Coding Fingerprints

After the acquisition process, the coding process typically compresses the image data to smaller storage sizes. One of the most common formats is the JPEG format, explained in Section 7.2.3, a container format that also defines a lossy compression of the image data. Lossy compression is when information is lost and cannot be restored during the decoding process. The information loss is optimized such that as much as possible of the information that a human cannot perceive is taken away, while as much as possible of the perceivable information will be stored.

The coding artifacts might be used for identifying the source device. Still, as the data input into this process is already digital, the identification would need to address the quality factor of the coded data, the quantization matrix, and the implementation of the coding algorithms. Limiting the possible parameters for the coding process will make it hard to identify the source device but might identify a class of devices or discard one device as the source.

Figure 7.5 Close-up of a JPEG photo showing the blocking effect from the DCT and quantization process. One such 8x8 block is highlighted in red at the top left corner of the image. The quality factor was set to 49 out of 100 in this photo (Photo by author).

As mentioned above, the coding will create artifacts in the photo, especially for lossy compression formats such as JPEG. The coding process has not been used for source device identification. Still, it has been used for assessing the authenticity of an image and used for establishing whether the image has been manipulated. It also serves as a method for ensuring that the photo has been processed in a consistent manner with the investigation hypotheses; otherwise, the hypothesis needs to be discarded.

The most notable image coding artifact in JPEG photos is the blocking effect of the DCT operating individually on 8×8-pixel blocks. Figure 7.5 shows an example of this blocking effect from a highly compressed JPEG image, and this blocking effect can be detected even for high jpeg quality settings.

Research by Fan and de Queiroz used a pretty simple method for finding whether such blocks were present in an image by comparing the gradient between border pixels between the blocks and comparing them to the gradient within a block (Fan & de Queiroz, 2003). If the gradients at the border regions are similar to those within the block, there is no proof of the existence of blocks. Still, if the gradients are consistently different between the border and the interior, there is proof of such compression blocks. After detecting the blocks, the authors use the histogram of neighbor pixel differences to find the likely quantization tables and the quality factor. In the case of the compression method using the quantization table, scaling this with the quality factor, the authors were able to correctly detect the quantization table used for scaling factor up to about 90.

Later research on the detection of the coding history of the photo is by considering three different types of encoders, namely transform-based encoders such as the DCT used in JPEG, sub-band coding, and differential image coding (Lin *et al.*, 2009). By assessing the type of encoder used, the type of equipment used for generating the photo can be established, even for a photo that has been printed. The knowledge about how many times and which settings have been used can also assess the photo's authenticity.

Double compressed photos are photos where the original photo is compressed, manipulated, and then compressed again. In this case, the artifacts from both coding processes can be present and detected. For JPEG images, two distinct types of double compression must be handled separately. The first type is *aligned double JPEG*, where the blocks from the two compression steps are aligned. The other type is *nonaligned double JPEG*, where the blocks from the two compression steps are not aligned.

The aligned double JPEG can be detected by inspecting the histogram of the DCT coefficient. Some have trained an SVMs to detect artifacts, and others have used analytical tools for detecting periodic artifacts in the histogram. Especially double peaks and missing centroids are telling of double compression.

Two different images double compressed by the same parameters yield two highly correlated images. The examined image will highly correlate with the right combination of compression settings by creating several test images with all different compression settings.

For nonaligned JPEG photos, the blocking effect can be exploited to find double compression evidence. Suppose the images have not been perfectly aligned. In that case, the original compression shows regular blocking artifacts, while the second compression would not exhibit such effects due to the edges caused by the first compression. Furthermore, misaligned blocks in the image can indicate tampering and recompression. Using an 8×8 blocking artifact filter over the whole picture, marking the areas where there is a blocking pattern and no pattern showing, manipulated areas can be found.

Machine learning, especially SVM, has been used for detecting double compressed images. Various sets of statistical data from the photo have been used to train SVMs to detect double compression artifacts.

One of the more novel techniques for recompression detection is the use of Benford's law. This law is more of an observation than a mathematical law, and it states that the most significant digit in a set of numbers follows a logarithmic curve. Mathematically speaking, the numbers follow Benford's law if the distribution of the most significant number is close to the distribution $P(d) = \log_{10}(1 + 1/d)$. The distribution was analyzed using the first digit of the DCT coefficients and compared to the expected result according to Benford's law (Milani *et al.*, 2012b). The method could identify up to four recompressions with over 94% true positive rate using an SVM classifier.

7.2.6 Editing Fingerprints

Image editing is the application of any processing to a digital image, and there are several types of editing that can be done in an image. One set of editing operations are operations aimed at improving the photo's aesthetics: adjusting light and colors, cropping to arrange the main objects in the photo, and correcting for imperfections in the sensor or lens. Other editing operations are more malicious in type, and information is added to, removed from, or changed to give the photo a different meaning. Forensically, the interest is most often in detecting the malicious operations or showing that only innocent operations have been done.

A paper by Piva, reviewing the photo forensic literature, classifies the editing operations into three categories (Piva, 2013):

1) Enhancements: editing operations such as histogram equalization, color modification, and contrast adjustment.
2) Geometric modification: editing operations such as rotation, cropping, zoom, and shearing.
3) Content modification: editing operations such as copy-move, cut-and-paste, and copy-and-paste.

Of the three types, the first is typically innocent operations, the second can be innocent or malicious, and the last is typically malicious.

The copy-move operation is, according to Piva, the most studied editing operation from a forensic perspective and is where a part of the photo is copied and pasted into another location to give hide something behind the pasted content or to duplicate something already in the photo. Another typical edit operation is the cut-and-paste operation, where parts of one image are pasted into another to add something not present in the original.

The examination of editing operations can be divided into two main methods: the signal-based methods and the scene-based methods. Signal-based methods examine how edit operations affect the contents of the image and look for fingerprints present in common edit operations. Scene-based methods look for irregularities caused by the edit operations in the scene itself, such as shadows, light, and other geometric features. These two main methods are complementary; an editing operation hard to detect on the signal level can be easier to detect at the scene level.

7.2.6.1 Copy-Move Detection

As it uses already existing parts of the image, the copy-move operations will have similar artifacts such as noise and color as the rest of the photo. An exhaustive search for cloned areas is often too expensive computationally to be feasible in images.

On the other hand, block matching can be used for simplifying the search. The photo is segmented into overlapping blocks, searching for similar connected blocks. A part of the image that has been copied will have similar blocks in the same relative positions and signify an area that has been copied. This can be simplified by extracting information from each block and comparing the relative positions. A diagonal line from corner to corner has been proposed in the literature, and others have used the block DCT coefficients as matching features. There have, in addition, been proposed various other transforms that are rotation invariant.

Another method is to use scale-invariant feature transform (SIFT) local features, where this method uses SIFT descriptors to find similarities in the photo (Huang *et al.*, 2018). This method has a drawback: it is challenging to select the correct matching strategy and adequately partition the image.

Among unsolved challenges in this topic is finding the original part among copies in the photo. This challenge is sometimes important, as it can tell why the editing has been done. Other open challenges are to increase the precision of the detection, to detect scaled or rotated versions of the original part, and to detect small, copied parts.

7.2.6.2 Cut-and-Paste Detection

Cut-and-paste operation detection methods cannot use the methods described above, as these are designed to find copies of the same image parts in other places in the photo. Splicing operations where the pasted content comes from another photo does, however, introduce other types of artifacts that can be detected. Most methods described for detecting spliced parts of a photo uses scene-based methods.

One way of detecting splices objects in a scene is to carefully analyze the illumination in the scene to detect irregularities to shadows and illumination of the objects in the photo. The direction of the light source on a three-dimensional scene can be hard to detect in a two-dimensional photo. Several assumptions are often made to ease finding the light source (Johnson & Farid, 2005).

The first simplification to the task is by assuming a Lambertian surface on the object under examination. A Lambertian surface is a surface that reflects the light isotropically, which means that the surface reflects light equally in all directions. Velvet is an example of an anisotropic surface, which means the reflection changes depending on the orientation of the surface.

The second simplification assumes that the examined surfaces have a constant reflectance, which means that the reflective properties are the same over the whole surface.

The last two simplifications are that the light source is a point source infinitely far away and that the angle between the surface and the light source is between 0 and 90°. The last two means that the light from the source illuminates the scene from the same angle and that the light source is in the front of the surfaces. With these simplifications, the location of a light source can be estimated by:

Equation 7.6: Location of a light source

$$I(x,y) = R\left(\vec{N}(x,y) \cdot \vec{L}\right) + A$$

where I is the image intensity at (x,y), R is the reflectance. $\vec{N}(x,y)$ is the three-dimensional vector that is orthogonal to the surface, $\vec{L}$ is the three-dimensional vector pointing toward the light source, and A is the constant ambient light.

By using local patches along the edges of the object instead of the whole edge, Johnson and Farid were able to relax the assumption about a constant reflectance over the entire surface (Johnson & Farid, 2005). In addition, the authors were able to extend the analysis to a local light source instead of a light source infinitely far away and for multiple light sources.

The shape and direction of shadows can be used to analyze the light source's direction, and both the use of geometric features and light intensities are used to find inconsistencies in the light source. The light source's color can also be estimated, and by analyzing the illumination color at several selected areas of the photo, inconsistencies in the illumination color can be detected.

Instead of using the attributes of the light source to find irregularities in the photo, the geometric properties of the perspective can also be used. The principal point in a photo is the point of the camera center on the image plane and is usually close to the center of the captured image. By analyzing the perspective projections of texture surfaces, this principal point can be estimated for several parts of the photo. Any irregularities to this estimated point thus indicate a part that has been spliced into the image.

Text in the photo is often regularly sized and can be used for detecting perspective irregularities in the photo. The text is usually shown on a planar surface, which makes comparing perspective geometry easier. If the perspective is inconsistent, the photo is likely to have been manipulated.

Motion blur can also be used for detecting inconsistencies in the photo, as the motion blur from movements in the source device during exposure will affect all objects in the photo. Stationary objects will all have the same amount and movement, and objects in motion will add the motion blur from the camera to the motion blur due to the object's movement. For this examination, the image gradients' spectral components can be used to detect inconsistencies.

The sizes of the spliced objects are hard to get exactly right, so this is also an artifact to examine. The height ratio of objects can be calculated without knowing the camera parameters, and a height ratio bigger or smaller than anticipated can indicate splicing.

7.2.6.3 Resampling Detection

Geometric operations such as scaling and rotation of a photo affect the pixel placements in the photo and lead to *resampling* to make the moved pixels into the pixel grid. This resampling process introduces some correlations that are detectable and depends on the resampling kernel that is used. Using an EM algorithm, as mentioned in Section 7.2.4.2, the interpolation kernel parameters can be estimated. The probability map of the probabilities for being correlated with the neighbor pixels is created. Transforming this map to the frequency domain shows periodic elements as spikes in the transformed map.

The Fourier transform of the second-order derivative of the pixels in the photo also shows a periodicity, in the same way as the method above. These two approaches give similar results, and it has been shown that these are related (Gallagher, 2005).

7.2.6.4 Enhancement Detection

Median filtering is an editing process that removes noise from the image by exchanging the value of a pixel with the median value of the pixels in the neighborhood. The median filtering process often leaves adjacent pixels that share the same value, a so-called *streaking artifact*. This can be detected by taking the difference between groups of two adjacent pixels and plotting the histogram of these differences. The detection rate of this simple method is quite good. Still, the performance will degrade rapidly with an increased level of lossy compression, as adjacent pixels will not retain their shared value.

Other detectable enhancements include contrast enhancements and histogram equalization. These operations are detected by analysis of the histogram of the pixel values, as the operations leave certain peaks and zeroes. Even though the analysis is simple, it is also effective and can detect areas containing splicing boundaries.

7.2.6.5 Seam Carving Detection

Seam carving is the process of automatic content-aware cropping, where the algorithm detects paths in the image that contain little content. These paths are then cropped and the image stitched together, creating an image where the content-less areas are removed, and the content-holding areas have been moved closer. Figure 7.6 shows two images where the first is the original, and the second has been seam-carved with the ImageMagick command: convert image.jpg -liquid-rescale 75 × 100% image-lr.jpg. To the left is the original image, and the image to the right has been cropped to 75%. The dark area to the left has been cropped, the distance between the trees on the horizon is smaller in the seam-carved image, and the branches on the tree to the right are also thinner.

Seam carving can be detected by building a Markov model over the co-occurrences of pixel values and values in the frequency domain, using machine learning to classify seam-carved images. Other methods to detect seam carving are to exploit the fact that missing low-content paths in the image indicates that an image already has undergone seam-carving, that the histogram distribution shows slight deviations, and to analyze the statistical moments of the wavelet transformed image.

Figure 7.6 Seam carving (Photo by author).

7.2.7 Deepfake Creation and Detection

A new forensic challenge that has become more relevant today is detecting synthetic images of people. As the synthetic rendering of human faces has reached the level of photorealism, the importance of detecting these images is also increasing. Generated photos might be used to create fake identification documents for crimes or can be used in astroturfing[3] campaigns by creating fake user accounts on social media.

The term *deepfake* describes deep learning for generating fake content in images, video, or audio. Deep learning is not a well-defined term, but it consists of machine learning methods using artificial neural networks as the basis for learning (Garg *et al.*, 2019).

7.2.7.1 Creation

Generated photorealistic images of faces can be generated using a generative adversarial network (GAN), which has two systems working together: a generator and a discriminator. The generator will produce the images, and the discriminator will classify the generator's output into the categories *generated* or *real*. More precisely, the generator updates the model it uses for generating images, while the discriminator will guess whether the images are from the model distribution or the data distribution (Goodfellow *et al.*, 2014). These two parts learn from each other. Each iteration improves the result until the generator reaches an acceptable level of realism or the generator and discriminator do not improve their results.

A team from Nvidia created an architecture for generating photorealistic images using a style-based GAN (Karras *et al.*, 2020). Examples of images produced with this architecture can be found on websites such as thispersondoesnotexist.com. The part of StyleGAN that distinguishes it from an ordinary GAN is in the generator, where a mapping network transforms the input to an intermediate format, where styles can be defined and fed to the synthesis network, generating the image.

There exist also other GANs for creating photorealistic images, such as cGAN, SRGAN, and StackGAN, among others, each with their enhancements to the original GAN architecture.

---

3 Astroturfing is to fake a grassroots movement using all available means to make other people think that there is massive support for the goal of the campaign.

7.2.7.2 Detection

Even though the human brain is capable of spotting artifacts in images that do not belong in a real scene or on a real person, detecting photorealistic images can be difficult. However, artifacts are often found from the synthesis process in generated images. They are invisible to the human eye but easy to spot using computational methods.

By analyzing an image in the frequency domain, Frank *et al.* were able to show a regular pattern in the generated images that were detectable with a linear ridge regression model in the frequency domain, resulting in a 100% detection rate (Frank *et al.*, 2020). The authors used DCT to transform the images to the frequency domain. In this domain, a pattern emerged and revealed the generated images, indicating that artifacts from the generator went undetected for the discriminator.

A photograph does tend to have frequency coefficients follow a $\dfrac{1}{f^{\alpha}}$ curve, with $\alpha \approx 2$ and f being the frequency along an axis, but the generated photos deviated from this. According to the authors' discussion, the artifacts may stem from the up-sampling when the GAN maps the data from the latent image space to the data space, from a vector of approximately 100 values to an image with 1024×1024 values. The GANs include methods for smoothing out these up-sampling artifacts from the spatial domain using nearest neighbor, bilinear, or other low-pass filters. Even with these methods, the artifacts are visible in the frequency domain and classified with a linear regression model on the log scaled DCT coefficients.

Figure 7.7 shows the average DCT coefficients for 2000 images created by StyleGAN2, using the Flickr dataset for training the GAN. The artifacts are not apparent to a human eye, but a pattern of squares is shown in this image, with the most noticeable artifact being the cross in the middle of the figure. To create this figure, the dataset from Nvidia

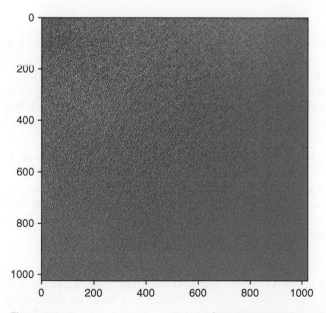

Figure 7.7 The averaged DCT coefficients for 2000 StyleGAN2-generated images. Note the spot in the middle, which is in the center of a barely noticeable brighter cross.

released with the original paper was used (Karras *et al.*, 2020), 2000 of these images were loaded in a Python program and processed. The values of the resulting figure are clipped at the 10 and 90 percentiles, so the DC values do not overshadow the rest of the frequency values.

Another approach to detecting generated photos is to use a neural network and train it on a set of images generated from various other systems. A convolutional neural network (CNN) has been reported to correctly classify all images where the training set and the test set are from the same generator (Wang *et al.*, 2020). The authors in this study also tested how general the classifier was by training the classifier on images from one generator, ProGAN, and using images from ten other generators in the test set. The results showed that the classifier correctly picked out the generated photos from other generators with an average precision of between 67 and 98%.

The study's authors also simulated editing operations by applying blur, JPEG compression, or both to the images before training the dataset. The generalization capabilities improved for almost all test datasets, with two exceptions: testing the classifier on the generators *SAN* and *DeepFake* showed that the average precision decreased from 93.6 to 53.7% when training with blurred images and testing on the SAN dataset, and from 98.2 to 66.3% when training with blurred and jpeg compressed images and testing on the DeepFake dataset. The authors discussed the reason for this decrement in precision; in the case of SAN, it might stem from the removal of high-frequency components by which the SAN-generated images can be detected.

The same study also showed that by including more classes from the training dataset, the generalization improved until about 16 classes were used. For more than 16 classes in the training set, the improvements to the generalization flattened out, and in some cases, decreased.

7.2.7.3 Challenges

The actual positive rate reported for the techniques used to detect generated faces can lead an investigator to place too much confidence in the result from the classifier. However, even with improving the robustness of the classifiers, the results can still be sensitive for crafted noise added to generated images. Carlini and Farid showed in their research that they could lower the detection rate down to zero in some cases by adding specially crafted noise to the generated images and testing this on the classifiers described above (Carlini & Farid, 2020).

The typical attack tricks the classifier into misclassifying synthetic images as real photos for an adversary. Also, misclassifying real photos as fakes is an attack that can be performed, albeit not as common as the misclassification of fake images as real. The study attacked the forensic classifiers for two different knowledge bases: a white-box approach, where the training set and the classifier's parameters are known, and a black-box approach, where the classifier's parameters are unknown.

7.2.7.4 White-Box Attacks

The attacker can compute the relation between the gradient input to the classifier and the output classification for the white-box attack. The attack was made by optimizing two factors, namely the classification precision and the amount of distortion needed to misclassify images.

The distortion minimizes the distortion by minimizing the *p*-norms of the added noise; in this case, 0-norm, 1-norm, 2-norm, and ∞-norm were used as measures. A *p*- norm is a distance measure and is given by the equation:

Equation 7.7: p-Norm distance measure

$$\|x\|_p = \left(\sum_{i=1}^{n} |x|^p \right)^{\frac{1}{p}}$$

where *x* is a vector, and *n* is the number of elements in the vector. A 2-norm is the Euclidian distance, and a ∞-norm is the maximum difference for any element in the vector: $\|x\|_\infty = \max_i |x_i|$.

The minimization is then given by finding the minimum amount of noise that still misclassifies the image:

Equation 7.8: Noise needed to misclassify an image

$$\operatorname*{argmin}_{\delta} \left(\|\delta\|_p \right), \text{s.t.} f\left(x + \delta \right) < \tau$$

By using Lagrangian relaxation, the hyperparameter, *c*, adjusts how much weight the optimization should put on the minimization of the noise about the misclassification rate:

Equation 7.9: Lagrangian relaxation

$$\operatorname*{argmin}_{\delta} \left(\|\delta\|_p + cf\left(x + \delta \right) \right)$$

For a parameter, *c*, the function is optimized using a gradient descent optimization, and the optimal *c* is found by a binary search in the parameter space.

This approach led to such small amounts of noise added that the quantization to 8 bits per color channel for storing the image as a PNG file resulted in removing the noise pattern. Instead, the distortion can be generated by flipping the least significant bits optimizing for the number of pixels that need to be changed before the image is misclassified.

The results for the number-of-pixel based attack on the classifier by Wang *et al.*, as described above, show that by changing 2% of the pixels, 71% of the generated images were classified as authentic images, and at 11% of the pixels changed, almost all generated images were misclassified. This shows that even though the classifier was designed to be robust, it is still sensitive to some input types.

The attack was, with some modifications, also successful against the method described above (Frank *et al.*, 2020). As the forensic method classifies not only into fake and real photos but also to which GAN was used to generate the image, the attack was modified such that it was only considered a success if a fake image was classified as a real image. This means that attacks misclassifying images detected by one GAN as made by another GAN as unsuccessful (as they were still detected as fake). By changing the least significant bit in 50% of the pixels, the detection rate fell to 0%.

The authors also tried three other types of white-box attacks, one with a simpler objective function, where the hyper-parameter from Equation 7.9 was removed, one where the objective function would optimize the noise pattern for all images instead of each image individually, and the last where the latent image space was modified to generate misclassified images. The latter method uses the attribute space the generator uses for generating the images. They searched for a global set of attributes that generated an image that will make the forensic classifier fail when added to the generator. Consequently, the images will be as photorealistic as the original images when the attack attributes are added.

The results showed that this method reduced the area under the curve (AUC) in a receiver operating characteristic (ROC) curve for the forensic classifier from 0.99 to 0.17. This means that for most of the tradeoffs between a high true positive versus a high false-positive rate, the false-positive rate will have to be higher than the true positive rate.

7.2.7.5 Black-Box Attacks

The white-box attacks are interesting, but more often, an attacker does not know about the forensic classifier's performance, and the relation between the inputs and outputs is unknown. In this regard, the forensic system is considered a black box, and the adversary does not have enough information to use a gradient-descent optimizer.

The black-box attacks are also successful, as two classifiers trained to perform the same tasks also tend to have similar vulnerabilities, as the decision-boundaries that the classifiers find also tend to be similar. Even by using a simpler forensic classifier to attack[4] as a stand-in classifier for the attack, which gave a much lower AUC than the original forensic classifier for detecting (0.85 as opposed to the original classifiers 0.96), the attack was able to generate images that were misclassified by the original. This reduction in classification success for the original method decreased from the AUC of 0.96 to 0.22.

The black-box attack was not as good as the white-box attack but shows that even without knowing about the forensic classifier's training parameters, it is possible to create noise patterns that, when added to a fake image, will reduce the success of the forensic classifier to less than chance. For a forensic examiner, it is important to be aware of the limitations of the tools available, under which circumstances they perform well, and under which circumstances they fail. When tested in isolation, a method with a good test score does not necessarily mean that the method is immune to an adversarial attack.

---

4 The original forensic classifier used ResNet-50, while this attack used ResNet-18.

7.2.7.6 Shallow Reconstruction

Another way to hide the artifacts from artificially generated images is to use a method called *shallow reconstruction* (Huang *et al.*, 2020). This method works in three stages, where the first is to create an attribute dictionary from real images, the second to map the fake images to the closest representation in the dictionary, and in the end, reconstruct the fake image using the dictionary. The term *shallow* refers to reconstructing the image by looking up and reconstructing from the dictionary in a projection from a single step.

The first step, creating the dictionary, is the step where machine learning methods are used to create a low-dimensional subspace of attributes. These attributes should capture the structures and artifacts of real images, and two methods for creating such a dictionary were tested: K-SVD (Aharon *et al.*, 2006) and principal component analysis (PCA) (Hotelling, 1935; Pearson, 1901). These two dictionary learning methods both find solutions to the problem:

Equation 7.10: Minimizing the error from dictionary reconstruction

$$\underset{\mathbf{D},\mathbf{X}}{\operatorname{argmin}} \| \mathbf{Y} - \mathbf{D}\mathbf{X} \|_F^2$$

where $\mathbf{D}$ is the learned dictionary, $\mathbf{X}$ is the coefficient matrix, and $\mathbf{Y}$ is the data, respectively. $\| \cdot \|_F$ denotes the Frobenius norm, a $L_{2,2}$-norm. K-SVD and PCA differ in their constraints for the objective function in this equation, where K-SVD is subject to $\forall i, \|\mathbf{x}_i\|_0 < K$, where $\| \cdot \|_0$ denotes the pseudo-norm measuring sparsity, and PCA is subject to $\mathbf{D}^T\mathbf{D} = \mathbf{I}$.

While K-SVD generates a sparse approximation matrix for $\mathbf{X}$, PCA creates a dense coefficient matrix. When performing a shallow reconstruction, the difference between the dictionaries is that while K-SVD results in an over-complete dictionary, the PCA results in an under-complete dictionary.

As the dictionary from the K-SVD method, $\mathbf{D}$, is overcomplete, a pursuit algorithm like the ortholinear matching pursuit (OMP) can find the sparse coefficient matrix for the image patch that is reconstructed, $\mathbf{x}$. For PCA, the representation matrix will be dense, and a least-square error solution can be used to obtain this. The contribution from the dimensions of the components can be adjusted to make the method more versatile. The reconstruction is simple for both methods with the dictionary and the image patch coefficient matrix and is given by $\hat{y} = \mathbf{D}\mathbf{x}$.

The experiments using this method showed that the misclassification rate went considerably up for images subjected to shallow reconstruction. The classifier that was used to classify images forensically was trained to classify fake and real photos and the architecture used for generating fake images. In all cases, the forensic classifier was fooled to classify a considerable amount of the generated images as real ones, and the PCA method for shallow reconstruction resulted in the worst forensic classifier performance, with close to 90% of the images created by ProGAN classified as real images. The best forensic classifier performance was for fake images created by MMDGAN, where the misclassification of the generated images as real ones increased from 0.11 to 45.94%.

The reason for the success of this method against forensic classifiers is that the dictionary is trained on real images and will pick up the notable artifacts that are present in real images. By reconstructing the representation of the fake images with the closest match from the dictionary of real images, artifacts that the forensic classifier has been trained to detect are thus removed with a high probability.

7.3   Video Forensics

Video forensics is in many ways like photo forensics, as a video can be thought of as being built up from a stack of images played back at a certain speed, like 30 or 60 frames per second. The capture process is similar, with the light passing filters, lenses, and a CFA before reaching the sensor, and the images must be demosaiced and encoded before being stored. However, a video does differ considerably from photographs after this point.

In addition to the visual part of the video, video forensics also includes audio forensics, as videos often contain an audio track. The inclusion of several synchronized media streams in a file also leads to examining the correlation between the streams to detect tampering with one or more of the streams.

Another difference between video and image forensics is the type of examinations done. There are huge economic interests in the movie industry and a considerable market for unlicensed copies of movies. There are also efforts to investigate who is behind the unlicensed copies. This means finding out where the copy has been made and attributing the unlicensed copies to the copyright infringers.

There are many good resources for a more in-depth look at video forensics. Milani *et al.* give a good overview of the topic in their survey paper and is a good starting point for diving deeper into the topic (Milani *et al.*, 2012a).

7.3.1   Video Process Flow

The video processing flow is like the photography process flow described in Section 7.2. Most of this process uses the same technology and process for imaging a scene to a digital storage media. The differences are primarily in that the acquisition process often captures sounds and the visual content and that the coding and compression step is better optimized for video content.

The coding and compression step for video codexes also uses several additional methods for compressing the resulting stream within acceptable quality parameters. Spatiotemporal prediction is used to predict the neighboring values, both in the spatial domain and between consecutive frames. In-loop deblocking filtering is done to smooth out sharp edges due to the blocking effect. It is called *in-loop* because it is done in the loop where the signal is fed back into the motion compensation algorithm and used for selecting the parameters for consecutively coding frames in the temporal domain.

The following subsections detail the types of artifacts generated during the video process and the differences between the video artifacts and the photographic artifacts.

7.3.2   Reproduction Detection

Detection of reproduction, or re-acquisition detection, detects some kind of new capture of a video displayed on a visual screen. Common ways of countering this are watermarking either video stream or onto a display screen.

In addition, there may also be found information about reproduction through:

- resizing and trapezoid adaption due to the post-processing of fitting the edges to the new format or size,
- sub-captions present,
- black edges due to format changes,
- differences compared to the original, if found, and
- source device identification, which we will discuss in the next section.

7.3.3 Source Device Identification

Source device identification is a vital examination task also in video forensics. As much of the process flow is the same as the photographic process, much of the theory and methods are the same as in image forensics. Some differences are mentioned in the subsections below.

7.3.3.1 Video Acquisition Artifacts

Like image forensics, the PRNU, CFA, and aberrations can detect the source device. As these methods are the same both in theoretic approach and implementation, please see the corresponding sections in Section 7.2.

Even with a vast number of images in which to measure the PRNU, the smaller image size and the vast compression ratio to the video stream make the noise more correlated to the coding parameters than to the PRNU. The original video PRNU found that a wavelet-based filter for denoising filters was the better choice for extracting the Gaussian noise of a given variance (Chen *et al.*, 2007).

Another filter to increase the visibility of the PRNU noise is to use a frequency domain filter to filter the frequencies corresponding to blocking and ringing artifacts at the block boundaries of compression blocks. Even using a better denoising filter, the number of frames needed to get a good detection rate increases with lower quality. For a video compressed to 4–6 Mb/s, about 40 seconds of video is required with the wavelet-based filter to get a reasonable detection rate. Still, about 80 seconds of video is needed if the video is compressed to 150 Kb/s.

7.3.3.2 Coding Artifacts

As videos are most often highly compressed, each device type usually has its specific setup of proprietary and public coding formats and settings to reach the compression levels needed for this compression. The set of video coding algorithms and settings makes the coding itself an artifact that can identify the device used to capture the video.

The possible coding algorithms and parameters used for videos are much more diverse than those used for images. As the file size for an uncompressed video is huge, aggressive compression algorithms are needed. Furthermore, the considerable correlation between pixels in the time domain makes the coding and compression very efficient in this domain.

Most of the video coding algorithms are based on the same principles as the JPEG standard. The frames are compressed by converting to a color space that allows for better compression before being split up in nonoverlapping blocks, transformed to the frequency domain, quantized, and entropy encoded. In addition, there are coding steps that compress the intra-frame correlations, blocks are predicted both spatially and temporally, and adaptive coding can be used to change settings and coding methods for different parts of the video.

The *blocks* of the video stream are not constant in videos, so the blocks themselves can reveal information about the coding method and settings used. It is mainly the block size that can vary. The edges of the blocks can be estimated by checking the local consistency of pixels or analyzing the frame in the frequency domain to find the peaks associated with the block boundaries.

The local consistency of pixels can be checked by comparing pixel differences between the pixels within a block from the pixels at block boundaries. A compression algorithm that uses blocks has been detected if there is a distribution difference between the differences at the boundary pixels and the ones within a block. Another approach is to analyze the frames in the frequency domain and detect the peaks in this domain that corresponds to the edges of the blocks in the spatial domain.

Modern codecs like H.264/AVC and HEVC use a deblocking filter that smooths the edges of the blocks. This filter is a challenge for the block detection methods described above. One way for finding the blocks was to scan for continuous horizontal and vertical intensity changes in the U channel of the YUV color space (Tou *et al.*, 2014). From these lines of continuous changes, a ladder structure was found to reveal the block size even for videos where the H.264/AVC deblocking filter had been used.

The quantization step will lose information, and this information loss can be detected. It can tell an investigator which type of quantization is used with the parameters and help identify the equipment used for the encoding. One way of detecting the quantization steps is to analyze the histogram of the transform coefficients, like the DCT coefficients for codexes using this transform. If the histogram has a comb-like structure, the quantization steps can be found by the distance between the peaks.

The motion prediction strategies used for the videos can also reveal information about which type of equipment has been used, especially the rate-distortion settings and the computational limits of the source device. The prediction step usually predicts motion vectors that express where the pixels in the video will move, and which vectors are used can also tell the type of source device. Many codexes use a block-based motion prediction method, such that a motion vector is estimated for each block.

To find the parameters for the motion prediction in H.264/AVC streams, Valenzise *et al.* (2010) propose to estimate the quantization parameter and the motion vector and find the set that minimizes the error. Their results showed that the estimated parameters were very good for high bitrate streams and performed well with lower bitrates.

Editing operations on compressed videos result in the video being decompressed, edited, and then recompressed. The recompression of an already compressed video stream can also be detected. While the last coding artifacts can be detected by the methods just described, it is also essential to detect the parameters used for the original video encoding and compression.

Section 7.2.5 describes how to detect double compression in images, and the same techniques can also be used for detecting recompression in videos, especially MPEG videos. There have also been efforts to detect such recompression in the High-Efficiency

Video Coding (HEVC) video codec, where the degradation of the image quality due to the double compression was detected (Jiang *et al.*, 2020). Many other techniques exist for video recompression detection, and many use statistical artifacts from the coding process to predict recompression.

7.3.3.3 Network Traces

As many video files have been streamed over a network before being saved, and most streaming protocols use the stateless User Datagram Protocol (UDP) as transport protocol, there is a nonnegligible probability for packet loss or packet reception after the playback video buffer already has played. The packet loss affects the stored video stream, leading to missing information in the video stream. Detecting this can be evidence of the video file source and the network quality between the streaming source and the receiving device.

The amount of information available for the investigator can be categorized into two categories: either the examiner has access to the source file or other versions of the streamed file, or only the video file under examination is available for the examiner. The latter can be described as a no-reference examination, as no reference video is available for the examination. As with other coding-related examinations, many techniques are based on modeling the statistical artifacts from the coding stage and compare with the examined video editing operations.

Editing operations are done for the same reasons in videos as in images. In the same way, as videos include more dimensions than images, the editing operations will also have to consider both the temporal aspect and the audio when editing. The causes for editing can also be classified into the same classes as for images, as described in Section 7.2.6, with the addition of audio modification and temporal modification. Temporal modification is like geometric modification but in the temporal instead of the spatial domain, where time is compressed or stretched out to either emphasize certain scenes or to change the impression of the contents, e.g., to let two unrelated events seem to be related by adjusting the time between them.

A video can be considered a consecutive set of images, and image forensic techniques for detecting editing operations can, in theory, be used for videos. In practice, the use of image forensic tools is impractical for several reasons (Milani *et al.*, 2012a): Image forensic tools are computationally demanding, replication and deletion of frames are undetectable, and the inter-frame relationship can contain artifacts after edit-operations.

7.3.3.4 Camera Artifacts

The artifacts from the photographic acquisition process can be used for detecting editing operations, such as checking the consistency of PRNU in the video frames. The PRNU can be compared by finding the correlation coefficient between a video frame and a reference frame, consecutive frames, or two frames without extracting the noise. If these three correlations coefficients are thresholded and the combination of these thresholded coefficients can detect editing operations such as frame insertion, frame replication, and copy-paste attacks.

Other camera artifacts such as the temporal correlation of residual noise and the analysis of the noise level function from photon shot noise have been explored. Still, the performance of these methods rapidly decreases with increasing compression. This decrease in performance makes the methods unreliable for ordinary MPEG-2 and H.264 compressed videos.

The performance of editing detection based on camera artifacts is limited by the high compression rates in videos. The higher compression, the more of the artifacts is hidden by the noise introduced with the compression of the video.

7.3.3.5 Coding Artifacts

The compression and coding step makes detecting the camera-based artifacts introduced by editing operations hard. However, the artifacts introduced during the coding step can also detect the aforementioned editing operations.

Intra-frame artifacts can be found in a single frame and are the same as the image forensic methods for detecting double compression. Looking at the inter-frame artifacts, the group of pictures (GOP) found in MPEG encoded videos will likely be changed after an editing operation that deletes or inserts frames in the video. The GOP is a set of frames that starts with an initial frame (I), and the following frames contain the updates from the initial frame and are called a P-frame if it uses data from previous frames to calculate the pixel values. In addition, there are the B-frames that can use both data from previous frames and data from the following frames for calculating the pixel values.

As the GOP is dependent on the scenes in the frames, a change in the number of frames will, with a high probability, also change the structure of the GOP, as it is highly sensitive to temporal changes in the scenes (Wang & Farid, 2006). The P-frames in a GOP are highly correlated to the initial I-frame in the original video. The reorganization of the GOPs will, with a high probability, transfer some frames from one GOP to the surrounding GOPs. This shift of frames in the GOP structures will extend to the remaining video. P-frames that have been moved to another GOP will have a weaker correlation with the initial I-frame in the new GOP. By using a Fourier transform over the motion-error values for each frame, the periodicity of these lower correlation P-frames can be detected.

Another coding artifact that can be used is the detection of inconsistencies in the de-interlaced videos from an interlaced source (Wang & Farid, 2007). A de-interlaced video will update every second row at each update, which means that the missing rows will be a combination of spatiotemporally adjacent rows. If the video has been edited, the assumption that the row is a combination of the adjacent rows will not hold, and editing operations can be detected. This method assumes that the editing operations did not recreate the interlacing and de-interlacing artifacts carefully.

An EM algorithm is used to determine the likelihood of belonging to these regions to find whether the values in a region fit into either of the two models: the de-interlaced model and a non-de-interlaced model. The values in a de-interlaced region will follow the distribution given by:

Equation 7.11: Expectation-maximization algorithm

$$F_{row}(x,y,t) = \sum_{i\in\{2j+1\}} \alpha_i F(x,y+i,t) + \sum_{i\in\{2j\}} \beta_i F(x,y+i,t+1) + n(x,y), j \in (Z)$$

where $F_{row}(x,y,t)$ is the de-interlaced value in a row, x,y,t is the spatiotemporal location, $F(x,y,z)$ is the values in the pixel in the frame, α and β are the parameters of the de-interlacing filter, and $n(x,y)$ is independent and identically distributed Gaussian noise.

The filter size is not necessarily bound, but in practice, the filter size, i is bound within ±3 rows. The values for a non-de-interlaced region are assumed to follow a uniform distribution. The EM algorithm will then assign a probability for each pixel to which category they belong.

7.3.3.6 Content Inconsistencies

Inconsistencies in the video content can also exist in the temporal domain between the frames in addition to the intra-frame inconsistencies found in images. It is mainly the motion used to detect editing operations in videos.

Inpainting is an editing technique that automatically replaces a missing part of an image by extending the surrounding texture. This technique leaves ghost shadows in the frames, and these artifacts can be detected by making an accumulated difference image (ADI) from one reference frame. The ADI is created by first creating a matrix the size of the image where all values are set to zero. The next step is calculating the difference from the reference photo for each consecutive frame, increasing the ADI matrix value by one if the difference is above a given threshold.

This method also detects movements, so the method works best if the removed area is a moving object or has not contained any moving objects. If this is the case, the still area will show on the ADI, indicating infill. This method will not indicate *when* the infill operation has been done, as the temporal location is compressed to only one frame.

Physical inconsistencies in the video can indicate editing operations, and one such physical trait to examine is the natural movement of objects as they move through a gravitational field. The paths of the objects can be modeled in three dimensions and projected to the two-dimensional image plane. Any deviation between the modeled movement and the actual movement either points to unaccounted forces acting on the object, adjustments to the timing of the video, changes to the number of frames (deletion or addition), or adding objects to the video.

7.3.3.7 Copy-Move Detection

With the increased availability of computing power and machine learning techniques, the old classification of intra-frame vs. inter-frame will have to be extended. Intra-frame detects changes within one image at a time, like image forensics. At the same time, inter-frame commonly has referred to the detection of copy from adjacent frames in a video stream, comparable to copy-paste detection with a reduced number of potential sources. We will now likely see copy-move and copy-paste detection move into an indexing area using machine learning and then utilize forensic tools on frames with objects close to each other in the indexing database. No such experiment has been found in the literature at the time of this writing.

7.4 Audio Forensics

Audio forensics is the forensic examination of recorded audio. Today's audio is recorded primarily to a digital medium; therefore, the focus of this chapter is on digital audio. An introduction to forensic audio analysis with examples can be found in Maher (2015), Zakariah (2018), and Maher (2018).

The investigative questions can fall into several categories:

— Authenticity and integrity of a recording
— Audio enhancements
— Transcoding identification
— Codec identification
— Double compression detection
— Speaker recognition
— Speech recognition
— Spatiotemporal identification
— Detect events in the recorded environment

7.4.1 Audio Fundamentals

To understand the forensic artifacts introduced by the audio recording process and the factors that affect the interpretation of the audio, we need to understand what sound is, the recording process, and how the human hears and interprets sound. The physical environment will shape the audio wave before it reaches the microphone. The recording equipment and the encoding process add artifacts to the stored audio signal. In addition, human hearing and cognition are prone to biases and interpretations.

7.4.1.1 Sound and Sound Pressure

A *sound* is a pressure wave traveling through a medium, which in most cases is air. However, the medium can be something else; a sonar detects a sound that has traveled through water, and a string telephone transmits the sound wave through a string. A pressure wave is most often visualized as a function of pressure over time, giving it the waveform commonly seen when representing sound waves.

The *sound pressure* is the difference between the peaks of a pressure wave and the ambient pressure, where a higher difference means a higher sound pressure. In acoustics, the sound pressure relative to a reference is used and is measured in decibels (dB). A Bel is the logarithm of an intensity divided by a reference intensity, and the relative intensity of a sound wave can thus be expressed in decibels using the following formula:

$$dB = 10\log\left(\frac{\text{Intensity}_1}{\text{Intensity}_0}\right) = 10\log\left(\frac{\text{Pressure}_1^2}{\text{Pressure}_0^2}\right) = 20\log\left(\frac{\text{Pressure}_1}{\text{Pressure}_0}\right)$$

In the equation, the reference pressure, Pressure_0, is usually set to $20\,\mu\text{Pa}$. This reference is close to the limit of human hearing. The sound pressure is measured as the root-mean-square (RMS) of the sound detected by a microphone, and a 0 dB audio signal is therefore at the threshold of what a human can hear.

7.4.1.2 Frequency

A wave oscillates between a high- and low-pressure point. The rate of this oscillation is the wave's frequency and is measured in Hertz (Hz). One Hz is the same as one oscillation per second and can have the unit Hz or s^{-1}; both have the same meaning. A frequency of 10 Hz is ten oscillations per second, and 3000 Hz is 3000 oscillations per second. SI prefixes are usually used for higher (or lower) frequencies, such that 3000 Hz can be named 3 kHz (kilohertz) instead.

Figure 7.8 shows a recording of human speech in three zoom levels. The first subfigure shows just over one second of the audio recording, the second shows about 20 ms, and the last shows about 300 μs. The last one also shows the discrete samples of the recording. They are approximately 23 ms apart, corresponding to a sampling rate of 44.1 kHz: $1/44{,}100 \approx 0.000023$ seconds. Other typical sampling rates are 48, 96, and 192 kHz, but 44.1 and 48 kHz are the most common in ordinary recordings.

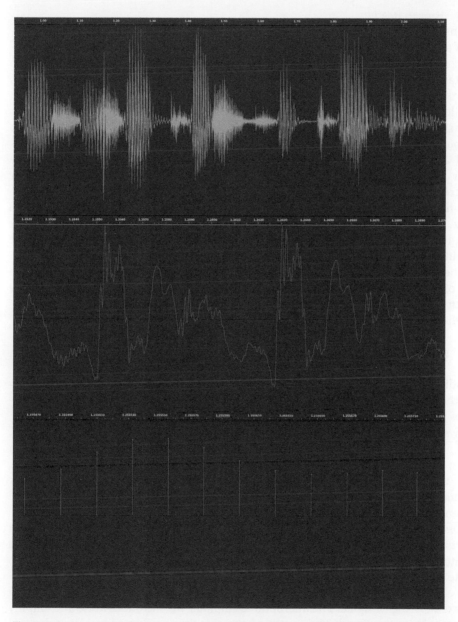

Figure 7.8 A waveform representation of an audio signal zoomed in at various levels, showing a second at the top, 20 ms in the middle, and individual samples at the bottom.

The Nyquist rate is the lowest sampling rate needed to reconstruct a signal and is twice the frequency of the highest frequency signal that is to be reconstructed. For example, for a sampling rate of 44.1 kHz, the highest frequency that can be reconstructed is 22.05 kHz. Most young people can hear frequencies up to 20 kHz, meaning any audible frequency can be reconstructed from a digital signal to an analog audio signal at this sampling rate.

7.4.1.3 Human Hearing

Humans have an outstanding capability of processing sounds, such as finding the direction of a sound source, focusing on one voice in a multi-voice environment, and interpreting spoken words in a noisy environment. In addition, other senses often help with these tasks, such as vision to help read on the lips, notice the speaker's body language, or see the response of other listeners.

The loudness of a sound is how loud a person perceives the sound. The loudness is highly subjective and will vary with both frequency and sound pressure. If listening to two sounds with the same sound pressure, the higher frequency sound is often perceived as louder than the lower frequency sound. The loudness is perceived more similar among varying frequencies at higher sound pressures than at lower pressures.

The human ear can only hear a limited frequency range, from approximately 20 to 20 kHz. The maximum audible frequency will decrease with age and is often 15–17 kHz for adults. The ear is most sensitive at 3 kHz, which is the frequency that contains most of the information in speech.

As the hearing is one of the essential tools for an audio forensic examiner, we must be aware of any limitations to the hearing. Regular measures of the hearing will reveal any hearing problems and makes us aware of any blind spots that would need extra attention when examining audio.

7.4.2 Digital Audio Recording Process

An audio wave in the air can be recorded by a microphone that converts the energy from the pressure wave into electrical signals. There are two major types of microphones used: dynamic and condenser microphones.

A *dynamic microphone* consists of a membrane that vibrates with the pressure wave. A coil attached to this membrane oscillates toward a stationary magnet, inducing an electric current that varies with the pressure wave. This microphone is inexpensive, sturdy, and can record high sound levels. However, the frequency response at higher frequencies is usually not the best, as the combined mass of the membrane and coil dampens the higher frequencies.

A *condenser microphone* has a better frequency response at higher frequencies and consists of a conducting backplate and a thin conducting sheet. This assembly is a capacitor, and it needs an electric potential between the plates, or *phantom power*, to operate.

The electric signal from a microphone is weak and needs to be amplified before being converted to a digital signal. The microphone amplifier will introduce noise and amplify existing noise to the signal. The amplified signal is then converted to a digital signal in an ADC. The ADC will also add noise to the signal, typically quantization noise.

The quantization noise is a type of noise that comes from the sampled real-numbered signal having a limited number of digital levels, which means that there will be a small error in the reproduction of the signal.

After the digitization in the ADC, the audio recording system can process the signal to improve the audio quality. Examples of this are high- or low-pass filters to remove unwanted high- or low-frequency noise, band-stop filters to filter out high-energy noise in small frequency regions, equalizer to emphasize or dampen frequency regions, or echo cancellation for removing feedback from speakers.

The last step in the recording process is encoding the audio signal, adding metadata, and storing it as a file. Many low-cost recorders will store the recording in a lossy mp3 format, while professional equipment often uses lossless encodings. A lossy encoding will discard information from the signal that humans are not good at hearing, just like the JPEG image format described earlier. A lossless format can be perfectly reconstructed into the original digital signal. Flac is an example of such a lossless audio format.

7.4.3   Authenticity Analysis

Many techniques can be used to establish the authenticity of a recording. If the recording contains a digital signature that can be verified, the authenticity is trivial to establish. Otherwise, the investigator must find proof of the authenticity by examining the audio recording for artifacts of its history. For example, editing operations such as cuts and splices can be detected, and room acoustics and background noise in the recording can be analyzed. In addition, metadata in the file can be analyzed to see if it is consistent with the recording. Evidence of manipulation will lead to a conclusion of an inauthentic recording. However, the lack of artifacts after manipulation is not proof of authenticity, as manipulation might go undetected.

7.4.4   Container Analysis

A container-based analysis will use elements like file structure, meta-data, and description of the audio file, not the audio data itself. Essential parts of container analysis are time stamps both on the file itself and registered in the meta-data of the file header. Other header information, like EXIF-information, is also a common source of information readily available with forensics tools. In addition, the completeness and forensic integrity checks using cryptographic hashes on all container parts will detect changes over time.

7.4.5   Content-Based Analysis

Edit operations can be detected by finding artifacts and inconsistencies in the recording that stems from these operations. As humans have a well-developed organ for analyzing sound, carefully listening to the recording can quickly determine whether there are sudden changes to the loudness, frequency spectrum, background noise, or sharp/artificial sounds, indicating a cut or splice in the recording.

Sudden changes can also be present in the waveform and spectrogram: Figure 7.9 shows a recording of speech where a word is spliced into the recording. The marker

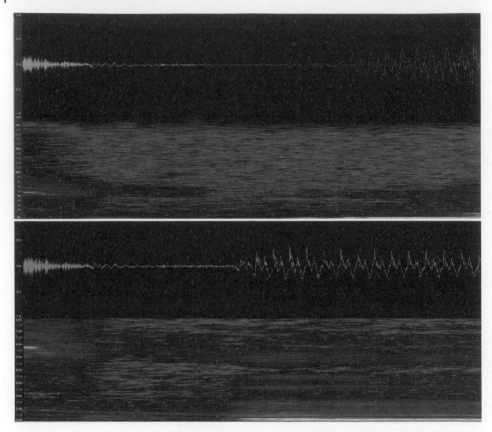

Figure 7.9 A word spliced into an audio recording in a combined picture of the waveform and spectrogram. The top shows the spliced recording, and the bottom shows the original recording.

shows the exact location, and the sudden change in the waveform and the spectrogram for the manipulated record is evident. There is also a high-frequency component at the splice that is not found anywhere else. The bottom figure shows the original recording.

Splicing can also introduce other artifacts, such as changes in reverb at different parts or different frequency contents due to another environment or recording equipment. A dynamic environment or moving the microphone between locations might also cause this, so the context of the changes must be considered.

7.4.6 Electric Network Frequency

A power grid operates on specific frequencies, either 50 or 60 Hz. For example, European countries use 50 Hz, while the USA uses 60 Hz. This electric network frequency (ENF) is found in all grid-connected devices, and the power lines radiate the signal like an antenna. Therefore, standalone electronic equipment can be affected by this ENF and be present as a hum at that frequency in a recording. For example, a 50 Hz hum will indicate that this has been recorded in a country running a 50 Hz ENF, and a recording claiming to be done in the USA with this hum will likely be a fake claim.

The ENF varies with the load in the grid. These variations will be picked up by audio recording equipment and be embedded in the audio track. The variability of the ENF can be used for establishing when a recording has been done by comparing the recorded ENF variations with data from the grid operator or electric power companies that are monitoring the health of the power grid. Another method for getting the ENF data is to record the ENF from the local power outlet, which requires resources to ensure that it follows regulations, monitors the equipment for failures, and regularly calibrates the equipment.

The examination of ENF signals can be considered signal forensics. However, as the signal sensing equipment is typically audio or video recorders, it also can be considered a part of audio forensics.

7.4.7 Audio Enhancements

In many investigations, audio recordings are essential evidence for establishing what has happened or been said. Unfortunately, the recording equipment in these cases has seldom been set up perfectly and might contain noise, low voices, or other sounds that drown out essential parts for the investigation. For example, a phone used for recording can have been put in a pocket, the recording can have been done in a car with the engine drowning out the voices, or it just might be a lousy quality microphone. In addition, listening to a noisy recording can be fatiguing for an investigator, and any unintelligible speech might be interpreted with a bias.

The *signal-to-noise ratio* (SNR) is the ratio between a signal and the noise floor in the recording. It measures how easily the signal can be interpreted; the lower SNR, the more difficult it is to pick out the signal from the noise. Conversely, a high SNR means that the signal is easy to pick out and conveys its encoded information without ambiguity.

Cleaning an audio recording most often refers to removing parts of the audio signal and factors that make it difficult to hear or understand the recording. Figure 7.10 is an example of a spectrogram of a person talking and a vacuum cleaner starting in the middle of the recording. The vacuum cleaner adds noise to the spectrogram, and the figure shows some high-energy frequency bands. A *notch-filter* is a narrow-band band-stop filter that can remove high-energy noise in a small frequency area. The figure to the right shows the effect of such a notch filter with a center frequency of 11,052 Hz.

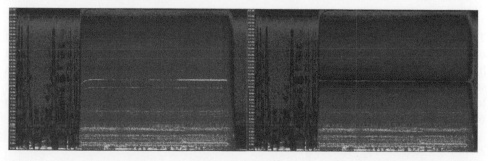

Figure 7.10 The spectrogram showing a vacuum cleaner starting while talking. The noise covers the whole frequency range of the recording. Running a notch filter to the left can remove high-energy, narrow-band noise.

High- and low-pass filters remove low and high-frequency parts of the spectrum, respectively. They let the chosen frequencies "pass" through the filter while blocking the rest. *Band-pass and band-stop filters* focus on specific frequency bands. *Equalizers* can also emphasize or deemphasize frequency areas instead of blocking them. Finally, some filters are more automatic, such as the noise reduction filter in the audio editor *Audacity*. This filter will analyze the noise in a silent part of the recording to learn the noise pattern and remove that pattern from the rest of the recording. Be aware that audio filters might remove parts of the signal that we want to keep.

7.4.8 Other Audio Forensic Methods

Many important forensic areas of audio files could be discussed and be addressed in greater detail here. We have selected a few below but recommend that the reader seeks other information sources for the following areas:

- *Speech recognition*. Detecting and potentially extracting speech from noisy recordings.
- *Speaker identification*. A highly active area of research for identifying people based on their characteristic speech patterns.
- *Environmental effects*. The detection of the events happening in the environment of the recording.
- *Acoustic environment signature*. The detection of the environment of an audio recording through the acoustic signature. For example, detection of a specific concert hall or conference room.

7.4.8.1 Codec Identification

Identification of the codec used to encode the audio is essential for forensic attempts in locating the origin, authenticity, or potential tampering information. Every time there is tampering with audio files, there has to be a sequence of decoding, change, and encoding taking place. Similarly, the use of multiple codecs can also take place when audio travels from older systems to newer systems, e.g., copper-based telephony to VoIP systems, or even from country to country.

Older detection techniques of codec identification are summarized in Yang *et al.* (n.d.).

7.4.8.2 Transcoding Identification

Transcoding is converting one audio encoding format to another. This could be from one file to another file or from one coding used in the creation, and another used when listening. Transcoding is often performed to save space in storage or transmission with some or potentially no loss of quality. The detection of transcoding from a lower bit rate to a higher bit rate may be performed by analysis of the high-frequency spectrum revealing identifying markers of the lower bit rate.

7.4.8.3 Double Compression Detection

Most compressed audio formats will be decompressed and then compressed again after being tampered with. The detection of double compression is, therefore, an area of active research that may reveal important information about the potential manipulation of the audio file. Yam *et al.* (2018) describe a potential method of detecting the compression history of MP3 files and thereby deriving an indicator of change.

7.4.8.4 Doppler Effect

When listening to a car's engine coming toward the listener, the engine's pitch, or frequency, will drop to a lower pitch as the car passes. This drop is the Doppler effect, caused by the compression of the pressure wave in the velocity direction, making the pitch higher. Conversely, the wave is stretched out behind the moving sound emitter, and the longer wavelength results in a lower pitch.

The speed of objects can be estimated in a recording by analyzing the Doppler effect. For simplicity, we can assume that the sound emitter is coming directly toward a stationary microphone. The frequency shift is constant as long as the speed is constant and will drop immediately when the sound emitter passes the microphone if the emitter travels directly toward the microphone. In that case, the frequency shift will be constant, but as the distance to the closest passing point increases, the frequency shift will slowly drop until the emitter has passed.

The equation for the frequency shift is dependent on the difference between the emitter and the microphone:

$$f = \left(\frac{c \pm v_m}{c \mp v_s} \right) f_0$$

where f is the heard frequency, c is the speed of sound, v_m and v_s are the speed of the microphone and the sound source, respectively. f_0 is the original frequency. The signs of the microphone and source velocities in the numerator and denominator are opposite.

For example, a police car driving at 100 km/h (22.78 m/s) is coming toward a stationary microphone recording. For simplicity, let's assume a static siren sound of 3 kHz. Using the formula, we see a shift in frequency of 8.81% higher when coming toward, and a change in frequency of 7.49% lower when passed. If the siren is originally at 3 kHz, the recorded sound will be 3.26 kHz and will drop to 2.78 kHz when the car is passing.

7.5 Summary

In this chapter, we have considered multimedia forensics and how multimedia can be used as evidence in cyber and digital investigations. We have reviewed various methods for metadata analysis, image forensics, video forensics, as well as audio forensics.

7.6 Exercises

1 What is metadata? Provide examples of metadata relevant to multimedia forensics and explain how it can be applied in investigations.

2 Describe the five classifications for forensic examination in image forensics.

3 Explain the process for deepfake detection and provide examples of how that can be used in investigations.

4   Use different recording equipment to record the same audio signal simultaneously. Analyze the recordings in an audio editor, such as Audacity. What is the difference between the signals? Are there any differences in the frequency domain between the recording equipment?

5   Edit a recording and see which effects this has on the audio signal. Is the manipulation detectable?

8

Educational Guide

Kyle Porter*

Department of Information Security and Communication Technology, Norwegian University of Science and Technology (NTNU), Gjøvik, Norway

This book assumes the reader has already read the previous book in the series, *Digital Forensics* by Årnes (2018), and builds upon this knowledge. Rather than burrowing deeper into the technical details of digital forensics, *Cyber Investigation* widens the scope of how investigations can be performed by focusing on crime in the cyber realm and the legal and technical processes of investigating it. Cyber investigations are still, however, very closely linked to the larger field of study of digital forensics.

This student guide's primary goal is to provide additional and practical resources to the reader that can further educate them beyond the intention of this book. Many of the listed resources typically have a more significant association with digital forensics, as the field is significantly more mature than cyber investigations. The technical tools for cyber forensics either come from traditional digital forensic areas or significantly overlap with them. We attempt to cover academic resources, nonacademic resources, tools, and datasets.

8.1 Academic Resources

This section attempts to encompass the primary resources used in the scientific field of digital and cyber forensics. We limit our resources to journals, conferences, and textbooks. Each subsection provides a list in alphabetical order of some of the most prolific journals, conferences, and books in the field.

Journals:

- *Forensic Science International: Digital Investigation* is likely the most influential journal in the field. Its content is usually broad but technical, as it accepts papers in about every subfield of digital forensics, with a newer interest in applied data science and machine learning. The journal also publishes the conference proceedings of the Digital Forensics Research Workshop (DFRWS) in special issues.

---

* Kyle Porter, Ph.D. is a Researcher at NTNU. Email: kyle.porter@ntnu.no.

Cyber Investigations: A Research Based Introduction for Advanced Studies, First Edition. Edited by André Årnes.
© 2023 John Wiley & Sons Ltd. Published 2023 by John Wiley & Sons Ltd.

- *IEEE Transactions on Information Forensics and Security*: IEEE is known for its quality, and this journal has a reasonably high impact factor. Its area of interest is quite broad compared to the other journals listed here, as they have articles from traditional computer forensics to cryptography and biometrics.
- *International Journal of Digital Crime and Forensics*: The IJDCF has general and technical digital forensics interests and articles on steganography and digital watermarking.
- *Journal of Digital Forensics, Security and Law*: An open-access journal that focuses on the legal and technical aspects of digital forensics.

Conferences and Organizations:

- *AnonBib:* While not a journal, AnonBib (https://www.freehaven.net/anonbib/) is an often-updated selection of the most seminal privacy-enhancing technology papers and deanonymization techniques. The papers are ordered chronologically and cached on the website, indexed by the author and the research area.
- *DFRWS*: The DFRWS conferences publish their proceedings as a special issue of FSI: Digital Investigations. There are now three primary conferences held by DFRWS each year, one in the United States (DFRWS USA), one held in Europe (DFRWS EU), and one held in the Asia-Pacific region (DFRWS APAC) as well. The conference is typically technical and covers a broad spectrum of digital forensics sub-disciplines.
- *IAPR*: The International Association for Pattern Recognition (IAPR) website describes itself as "an international association of nonprofit, scientific, or professional organizations (being national, multinational, or international in scope) concerned with pattern recognition, computer vision, and image processing in a broad sense." (IAPR, 2022) They hold many conferences that include topics related to digital forensics. Still, their largest focus seems to be applying cutting-edge machine learning methods to solve these problems. The reader's topics of interest may be handwriting recognition, biometrics, and computer vision.
- *IFIP WG 11.9:* The International Federation for Information Processing (IFIP) is the overarching organization that runs the Advances in Digital Forensics conferences and publishes the proceedings as a book. Papers here are often technical, but highly theoretical or legal work is encouraged.
- OSDFCon: A practically oriented conference, the Open-Source Digital Forensics Conference focuses on open-source digital forensics tools by giving workshops and talks related to them and even offers Autopsy training. The conference also hosts the annual Autopsy Module Development Contest, where the top three winners can earn up to USD 1000.
- PETS: The Privacy Enhancing Technologies Symposium focuses on researching anonymization networks such as TOR, as well as deanonymization techniques.

Textbooks:

- *The Art of Memory Forensics*: An authoritative technical book on memory forensics intended for law enforcement and researchers alike. This book is the complete package covering the basics and foundations of memory forensics, applying knowledge with the Volatility Framework, and advanced chapters on Windows, Linux, and Mac memory forensics.

- *File System Forensic Analysis* (Carrier, 2005): Another classic for digital forensics is the go-to source for file system forensics. Complete with a theoretical model on categorizing file systems, Brian Carrier also performs deep dives on the technical aspects of various file systems, including FAT, NTFS, and Ext2/3. Despite being more than 15 years old, this book is still highly relevant.

8.2 Professional and Training Organizations

We should note that unlike many of the other references in this chapter, training for the following organizations is not free of charge and very expensive in most cases. Despite this, such training and certifications are valuable.

- CISA/CISM: The Certified Information Systems Auditor certification and Certified Information Security Manager certifications are offered by the international organization, Information Systems Audit Control Associations (ISACA). According to their website, members work in more than 180 countries (ISACA, 2022), with various roles and levels of experience.
- CISSP: The Certified Information Systems Security Professional certification has members worldwide. However, over half of them are in the United States. It is approved by the United States Department of Defense. The CISSP certification is more technology-focused than the CISA or CISM certifications.
- *European Network of Forensic Science Institutes (*ENFSI*)*: One of the few uniquely European organizations, ENFSI is composed of a large set of working groups, all under the umbrella of general forensic science. The Forensic Information Technology working group is concerned with computer data analysis and Internet investigations.
- *Information Assurance Certification Review Board (*IACRB*)*: The certification body offers 13 job-specific certifications, including computer forensics examiner, mobile forensics examiner, threat hunting professional, and more.
- *International Association of Computer Investigative Specialists (*IACIS*)*: IACIS is unique from the previously listed organizations since it focuses on computer forensics rather than information security. The Certified Forensic Computer Examiner certificate is awarded upon completing their certification program.
- *SANS*: SANS is a US-based organization that provides training and certification for various fields in information security. Areas of training and certification include forensics, network security, malware analysis, and others. Their certification training is through Global Information Assurance Certification (GIAC), and trainees can earn a certification in several digital forensics subfields.
- *Tool specific Training:* It is quite common for the vendors of various forensics tools to provide training for their tools. The courses usually cost quite a lot of money, but they also frequently offer free webinars. For example, Basis Technology (Autopsy), Maltego, and MSAB (XRY) are vendors offering training programs.

8.3   Nonacademic Online Resources

Digital and cyber forensics are moving targets, and thus it can be tough to stay up to date. To help stay updated, we provide nonacademic digital forensic sources, such as blogs, forums, and YouTube Channels.

- The granddaddy for online digital forensics news and community is Forensic Focus. Constantly updated with opinion pieces, research results, and vendor plugs, this website is likely the most popular digital forensics website on the Internet. It would be remiss of us not to mention the Forensic Focus Forums, an active forum containing nearly decades of conversations by forensic professionals and newbies.
  - https://forensicfocus.com/
- The best aggregator for forensic news is likely thisweekin4n6.com (This Week in 4n6) by Phill Moore. Most weeks, the website provides a list of new resources divided into their respective topics, such as malware, software updates, forensic analysis, and threat intelligence.
  - https://thisweekin4n6.com/
- For very practical and hands-on learning of Digital Forensics and Incident Response content, see Richard Davis' 13Cubed and Joshua I James' DFIR. Science YouTube Channels. Their videos explain technology and concepts and also demonstrate the technology.
  - https://www.youtube.com/c/13cubed
  - https://www.youtube.com/user/dforensics
- A popular digital forensics podcast is David Cowen's "Forensic Lunch." The show typically features guests from the digital forensics community or runs tutorials.
  - https://www.youtube.com/user/LearnForensics

8.4   Tools

Tools for use in Digital and Cyber Investigation are typically either freely available or quite expensive. Here, we emphasize some of our favorite free-to-download tools for the various digital forensics branches, which can be applied in cyber investigations. Forensicswiki.xyz is a good resource for discovering such tools, but the site does appear to be less frequently updated than it used to.

A commonly used integrated kit of tools for forensics, penetration testing, security research, and reverse engineering is Kali, where many of the other tools mentioned in this section can be found. For more information about Kali how to install, use and download, see:

- https://www.kali.org/

8.4.1   Disk Analysis Tools

The go-to disk analysis tool for professionals and researchers alike for disk-based forensics is The Sleuth Kit (TSK). This command-line tool can display the partition layout of the system, provide details of the file system parameters for a given partition, retrieve file metadata content, and much more. Its GUI counterpart, Autopsy, is more user-friendly and essentially acts as a typical digital forensic suite. Autopsy has an active development community because Autopsy can be fitted with user-created modules written in Python (technically Jython). For example, these modules may allow the ingestion of previously unsupported data formats or even apply face recognition to images found on the disk. There is an annual competition to create these modules, run by the OSDFCon. For more information, see:

- https://github.com/sleuthkit/autopsy_addon_modules
- https://www.osdfcon.org/
- https://www.sleuthkit.org/

Another useful free forensic suite is the FTK imager, which supports many of the functionalities of Autopsy but also allows exporting disk images to raw files, which can be handy if you are only provided with an E01 file.

- Found at: https://accessdata.com/product-download/ftk-imager-version-4-5

8.4.2 Memory Analysis Tools

The first step for memory analysis is to acquire the data. Recall, this needs to be done with care as not to make any significant changes in memory. A few tools which can facilitate this task are as follows:

- FTK Imager (see link above)
- https://github.com/504ensicslabs/lime
- https://github.com/carmaa/inception

For performing more sophisticated memory analysis, Redline and Volatility are standard tools. Both tools can view running processes, where Redline uses a GUI and Volatility is command-line based. An additional benefit to Volatility is that it supports third-party plugins written in Python.

- https://www.fireeye.com/services/freeware/redline.html
- https://www.volatilityfoundation.org/

8.4.3 Network Analysis Tools

Wireshark is likely the most popular network analysis tool, and for good reasons. It can be used to sniff and save network data, enabling insight into packet metadata of all packets going in and out of your network or computer. If the payload of the packets is unencrypted, then the payload can be read too.

- https://www.wireshark.org/

Two common intrusion detection systems are Zeek (formerly Bro) and Snort. Snort is an intrusion prevention system as well, where it uses easy to construct "Snort rules" as signatures that can potentially block traffic after performing online deep packet inspection. Zeek acts quite differently as an IDS, as it records network activity and stores it in a fashion that can be effectively searched, thus making for more accessible network analysis.

- https://www.snort.org/
- https://zeek.org/

8.4.4 Open-Source Intelligence Tools

Visualization and correlation of open-source intelligence can significantly enhance the analysis phase of the digital forensics process, and a tool that supports this is Maltego. The tool's purpose is to discover links and relationships between entities that can be found through open sources. You can provide the tool with an entity such as a domain name, alias, document, and phone number. You can then run scripts that perform searches for related entities via open or Maltego partner sources. Maltego represents the entities as nodes in a directional graph.

- https://www.maltego.com/

The i2 Analyst's Notebook is a popular tool used by industry and has "multidimensional visual analysis capabilities" to discover latent networks or connections (N. Harris Computer Corporation, 2022). Unlike Maltego, there is no free community edition, and the cost for a license is quite steep.

- https://i2group.com/

WiGLE is a project that collects information on wireless networks worldwide and provides their locations, SSID, and BSSID identifiers in an interactive map. As of 18 January 2022, they have 11,816,441,771 unique Wi-Fi locations in their database (https://wigle.net/stats). Users can contribute to the database by submitting their discovered hotspots.

- https://wigle.net/

Another category of tool that should not be overlooked is social media APIs. These APIs allow users to interact with a given social network with their software. This can allow the performance of automated tasks, such as recording new and historical user comments. As most of the APIs are provided to the public by the social media company, their use typically comes with several rules, such as a maximum number of requests per minute or limitations on the extraction of historical data. Some popular social media APIs that can be used for OSINT are given as follows:

- https://www.reddit.com/dev/api/
  - https://praw.readthedocs.io/en/latest/
- https://developer.twitter.com/en
  - https://tweepy.readthedocs.io/

As mentioned in Chapter 4, Apache Tika is a valuable tool for extracting text and metadata content from PDFs and Office files. While not strictly a tool for open-source intelligence, it greatly complements the automated processing of documents. Furthermore, there is a Python port, which can easily be complemented with machine learning tools.

- https://tika.apache.org/
  - https://pypi.org/project/tika/

8.4.5 Machine Learning

On the occasions in which machine learning can be used in cyber investigations, we suggest using Python and Weka.

Python is not a *tool per se*, but a scripting language that is easy to use, extremely flexible, and has an extensive catalog of machine learning libraries that have been developed for it. A popular and easy-to-use library for machine learning is scikit-learn. It has modules that perform the most basic machine learning tasks such as classification, clustering, regression, dimensionality reduction, and more. For more heavy-duty tasks, such as image or speech recognition, we suggest libraries that can effectively use deep learning, such as PyTorch, TensorFlow, or Keras. Perhaps the easiest way to start is to install the Anaconda Python distribution, which is designed specifically for data science and machine learning, as it automatically installs many packages that support these tasks. Additionally, Python has extensive support for visualizing the machine learning analysis results. However, Python is slow, so the user may want to look into a lower-level language for computationally intensive cases, such as C++.

- https://www.anaconda.com/
- https://scikit-learn.org/stable/
- https://pytorch.org/
- https://www.tensorflow.org/
- https://keras.io/

Weka is a popular machine learning workbench that has often been used in digital forensics research and provides the user a GUI and the core machine learning abilities such as classification, clustering, feature selection, etc. Weka has been used by Flaglien (2010) as a tool to conduct identification of malware across multiple computers and Clemens (2015) to classify essential aspects of code.

- https://www.cs.waikato.ac.nz/ml/weka/

A valuable resource to apply Natural Language Processing to human-written text is Huggingface's *Transformers* library for Python:

- https://github.com/huggingface/transformers
- https://huggingface.co/transformers/

The library allows for access to many different state-of-the-art pretrained language models of various languages. The library also provides for relatively easy fine-tuning of the models to enhance the performance of varying language tasks such as Named Entity Recognition and Sentiment Analysis.

Anonymity tools and related research information:

- AnonBib – https://www.freehaven.net/anonbib/
- Tor Browser – https://www.torproject.org/
- Tails – https://tails.boum.org/
- Qubes OS – https://www.qubes-os.org/
- JohnDonym – https://anonymous-proxy-servers.net/
- Stem library – https://stem.torproject.org/

8.5   Corpora and Data Sets

The field of digital forensics has come quite far in providing data for researchers to test their tools on. Below, we provide some links that host many useful datasets.

Digital Corpora is likely the most popular website for digital forensics datasets. Primarily consisting of realistic synthetic disk images, they also provide network packet dumps, memory dumps, and a massive corpus of freely distributed files. The diversity and size of the disk images provided on the site are unparalleled. There are disk images containing a variety of file systems, either extracted from a desktop computer or mobile phone.

- https://digitalcorpora.org/

Computer Forensic Reference Data Sets is a NIST-run website that contains a large set of relatively small but task-oriented files for digital forensics. Example tasks include file carving and deleted file recovery, string searching, Unicode string search, and searching for evidence for images of rhinos. Another benefit of the website is that the same disk images are provided in various file systems (FAT, XFAT, NTFS, and ExtX).

- https://www.cfreds.nist.gov/

Brian Carrier's Digital Forensics Tool Testing Images is a rather old but still relevant website hosting forensic datasets. The use cases are like NIST's, but the images are even smaller, and the website provides the solutions to the tasks.

- http://dftt.sourceforge.net/

Grajeda et al. (2017) surveyed what datasets were available for digital forensics, and links to their results are hosted by the Datasets for Cyber Forensics website. This website hosts the most diverse datasets of all the sources referenced thus far, from chat logs, malware, email datasets, leaked passwords, pictures, network traffic, disk images, memory images, etc. It also links the popular Enron email dataset (Cohen, 2015).

- https://datasets.fbreitinger.de/datasets/
- https://www.cs.cmu.edu/~./enron/

The Volatility project also hosts a number of memory images free for download.

- https://github.com/volatilityfoundation/volatility/wiki/Memory-Samples

Likewise, the Wireshark project hosts a vast number of capture files free for download.

- https://wiki.wireshark.org/SampleCaptures

8.6 Summary

While we have attempted to aggregate many of the external resources that may be useful for a cyber or digital forensics student, we have undoubtedly missed important sources. We hope we have provided you with a springboard to continue your theoretical and practical education despite this acknowledgment. On the other hand, you may be overwhelmed with the number of different subjects and resources we have provided. To this, we suggest exploring the resources that you find interesting rather than studying them from end to end. We wish you the best of luck with your future endeavors.

References

Abbott, T. G., Lai, K. J., Lieberman, M. R., & Price, E. C. (2007, June). Browser-Based Attacks on Tor. *Proceedings of the Seventh Privacy Enhancing Technologies Symposium (PETS 2007)*. Ottawa: Springer.

Ablon, L. (2018, March 15). *Data Thieves - The Motivations of Cyber Threat Actors and Their Use and Monetization of Stolen Data*. Retrieved from rand.org: https://www.rand.org/pubs/testimonies/CT490.html

ACPO (2005). *Core Investigative Doctrine*. London: ACPO.

Aharon, M., Elad, M., & Bruckstein, A. (2006, November). K-SVD: An Algorithm for Designing Overcomplete Dictionaries for Sparse Representation. *IEEE Transactions on Signal Processing 54*, 4311–4322.

Aitken, C., Roberts, P., & Jackson, G. (2010). *Fundamentals of Probability and Statistical Evidence in Criminal Proceedings: Guidance for Judges, Lawyers, Forensic Scientists and Expert Witnesses*. London: RSS.

Amoore, L. (2020). *Cloud Ethics - Algorithms and the Attributes of Ourselves and Others*. Durham and London: Duke University Press.

Anderson, T., Schum, D., & Twining, W. (2005). *Analysis of Evidence*, 2nd ed. Cambridge: Cambridge University Press.

Antonakakis, M., April, T., Bailey, M., Bernhard, M., Bursztein, E., Cochran, J., Halderman, A., Kallitsis, M., Kumar, D., Lever, C., Ma, Z., Mason, J., Menscher, D., Seaman, C., Sullivan, N., Thomas, K., & Zhou, Y. (2017). Understanding the Mirai Botnet. *Proceedings of the 26th USENIX Security Symposium*. Retrieved from https://www.usenix.org/conference/usenixsecurity17/technical-sessions/presentation/antonakakis

Ashton, K. (2009). That 'Internet of Things' Thing. *RFiD Journal 22* (7), 97–114.

Ask, K., & Granhag, P. A. (2005). Motivational Sources of Confirmation Bias in Criminal Investigations: The Need for Cognitive Closure. *Journal of Investigative Psychology and Offender Profiling 2* (1), 43–63.

Atzei, N., Bartoletti, M., Lande, S., & Zunino, R. (2018). A Formal Model of Bitcoin Transactions. *International Conference on Financial Cryptography and Data Security* (pp. 541–560).

Atzori, L., Iera, A., & Morabito, G. (2014). From "Smart Objects" to "Social Objects": The Next Evolutionary Step of the Internet of Things. *IEEE Communications Magazine 52* (1), 97–105.

Avizienis, A., Laprie, J. C., Randell, B., & Landwehr, C. (2004). Basic concepts and taxonomy of dependable and secure computing. *IEEE Transactions on Dependable and Secure Computing 1* (1), 11–33. https://doi.org/10.1109/TDSC.2004.2.

Azhar, M. A., & Bate, S. B. (2019). Recovery of Forensic Artefacts from a Smart Home IoT Ecosystem. *CYBER 2019: The Fourth International Conference on Cyber-Technologies and Cyber-Systems.*

Back, A. (1997, March 28). *A Partial Hash Collision Based Postage Scheme.* Retrieved from hashcash.org: http://www.hashcash.org/papers/announce.txt

Badenhop, C. W., Ramsey, B. W., Mullins, B. E., & Mailloux, L. O. (2016). Extraction and Analysis of Non-Volatile Memory of the ZW0301 Module, a Z-Wave Transceiver. *Digital Investigation 17*, 14–27. https://doi.org/10.1016/j.diin.2016.02.002.

Barbulescu v Romania, Application no. 61496/08 (European Court of Human Rights (ECtHR) September 5, 2017).

Bayram, S., Sencar, S., & Memon, N. (2006). Identifying Digital Cameras Using CFA Interpolation. In J. P. Sandvik (Ed.), *International Federation for Information Processing,* vol. 222, ch 23 (pp. 289–299). Retrieved from https://www.researchgate.net/publication/221352761_Identifying_Digital_Cameras_Using_CFA_Interpolation.

Bazán-Vera, W., Bermeo-Almeida, O., Samaniego-Cobo, T., Alarcon-Salvatierra, A., Rodríguez-Méndez, A., & Bazán-Vera, V. (2017). The Current State and Effects of Agromatic: A Systematic Literature Review. In *Communications in Computer and Information Science,* vol. 749 (pp. 269–281). Berlin: Springer Verlag.

BBC News. (2008, February 26). *Pakistan Lifts the Ban on YouTube.* BBC News. Retrieved from http://news.bbc.co.uk/1/hi/technology/7262071.stm

Berthold, O., Federrath, H., & Köpsell, S. (2000a, July). Web MIXes: A System for Anonymous and Unobservable Internet Access. In H. Federrath (Ed.), *Proceedings of Designing Privacy Enhancing Technologies: Workshop on Design Issues in Anonymity and Unobservability,* LNCS 2009 (pp. 115–129). Springer-Verlag. Retrieved from https://link.springer.com/chapter/10.1007/3-540-44702-4_7.

Berthold, O., Pfitzmann, A., & Standtke, R. (2000b, July). The Disadvantages of Free MIX Routes and How to Overcome Them. In H. Federrath (Ed.), *Proceedings of Designing Privacy Enhancing Technologies: Workshop on Design Issues in Anonymity and Unobservability,* LNCS 2009 (pp. 30–45). Springer-Verlag.

Bjelland, P. (2018). Internet Forensics. In A. Årnes (Ed.), *Digital Forensics.* Hoboken, NJ: John Wiley & Sons, Ltd.

Bjerknes, O. T., & Fahsing, I. A. (2018). *Etterforskning. Prinsipper, Metoder og Praksis.* Bergen: Fagbokforlaget.

Bonomi, F., Milito, R., Zhu, J., & Addepalli, S. (2012). Fog Computing and Its Role in the Internet of Things. *MCC'12 - Proceedings of the 1st ACM Mobile Cloud Computing Workshop* (pp. 13–15).

Bouchaud, F., Grimaud, G., & Vantroys, T. (2018). IoT Forensic: Identification and Classification of Evidence in Criminal Investigations. *Proceedings of International Conference on Availability, Reliability and Security.* https://doi.org/10.1145/3230833.3233257

Brandom, R. (2013, December 18). *FBI Agents Tracked Harvard Bomb Threats Despite Tor.* The Verge. Retrieved from https://www.theverge.com/2013/12/18/5224130/fbi-agents-tracked-harvard-bomb-threats-across-tor

Buchanan, T. R. (2015). Attributing Cyber Attacks. *Journal of Strategic Studies 38*, 4–37.

Bult, K., Burstein, A., Chang, D., Dong, M., Fielding, M., Kruglick, E., Ho, J., Lin, F., Lin, T., Kaiser, W., Marcy, H., Mukai, R., Nelson, P., Newburg, F., Pister, K., Pottie, G., Sanchez, H., Stafsudd, O., Tan, K., Xue, S., & Yao, J. (1996). Low Power Systems for Wireless Microsensors. *Proceedings of 1996 International Symposium on Low Power Electronics and Design.* 10.1109/lpe.1996.542724

Burgess, M. (2018, April 20). *From Fitbits to PlayStations, the Justice System is Drowning in Digital Evidence.* Wired.com, Retrieved from https://www.wired.co.uk/article/uk-police-courts-data-justice-trials-digital-evidence-rape-cases-cps

Cambridge Dictionary. (2022, January 29). *Cybercrime.* Cambridge Dictionary. Retrieved from https://dictionary.cambridge.org/dictionary/english/cybercrime

Camenzind v Switzerland, 136/1996/755/954 (European Court of Human Rights (ECtHR) December 16, 1997).

Canadian Centre for Cyber Security. (2018, December 6). *Cyber Threat and Cyber Threat Actors.* cyber.gc.ca, Retrieved from https://cyber.gc.ca/en/guidance/cyber-threat-and-cyber-threat-actors

Carlini, N., & Farid, H. (2020). Evading Deepfake-Image Detectors with White- and Blackbox Attacks. *IEEE Computer Society Conference on Computer Vision and Pattern Recognition Workshop* (pp. 2804–2813). IEEE.

Carrier, B. (2005). *File System Forensic Analysis.* Upper Saddle River, NJ: Addison Wesley Professional.

Carrier, B. D., & Spafford, E. H. (2004, August 11–13). An Event-Based Digital Forensics Investigation Process. *Digital Forensics Research Workshop (DFRWS).* USA.

Casey, E. (2002). Error, Uncertainty and Loss in Digital Forensics. *International Journal of Digital Evidence 1* (2). Retrieved from https://dblp.uni-trier.de/rec/journals/ijde/Casey02.html?view=bibtex.

Casey, E. (2011). *Digital Evidence and Computer Crime: Forensic Science, Computers and the Internet.* Amsterdam: Elsevier.

Casey, E. (2016). Differentiating the Phases of Digital Investigations. *Digital Investigation 19*, A1–A3.

Casey, E. (2020). Standardization of Forming and Expressing Preliminary Evaluative Opinions on Digital Evidence. *Forensic Science International: Digital Investigation 32*, 200888.

Center for Cyber Security (CIS). (2019, January 13). *Cybersecurity Spotlight – Cyber Threat Actors.* cissecurity.org, Retrieved from https://www.cisecurity.org/spotlight/cybersecurity-spotlight-cyber-threat-actors/

Center for Internet Security. (2022, January 23). *What is Cyber Threat Intelligence?* Center for Internet Security (CIS). Retrieved from https://www.cisecurity.org/what-is-cyber-threat-intelligence/

Charman, S. D., Carbone, J., Kekessie, S., & Villalba, D. K. (2016). Evidence Evaluation and Evidence Integration in Legal Decision-Making: Order of Evidence Presentation as a Moderator of Context Effects. *Applied Cognitive Psychology 30*, 214–115.

Chaum, D. L. (1981, February). Untraceable Electronic Mail, Return Addresses, and Digital Pseudonyms. *Communications of the ACM 24*, 84–90.

Chaum, D. L. (1988, January). The Dining Cryptographers Problem: Unconditional Sender and Recipient Untraceability. *Journal of Cryptology 1*, 65–75.

Chen, M., Fridrich, J., Goljan, M., & Jan, L. (2007). Source Digital Camcorder Identification Using Sensor Photo Response Non-Uniformity. In J. P. Sandvik (Ed.), *Security, Steganography and Watermarking of Multimedia Contents IX*, vol. 6505 (p. 65051G. Retrieved from https://www.researchgate.net/publication/228769530_Source_digital_camcorder_identification_using_sensor_photo_response_non-uniformity_-_art_no_65051G.

Cheswick, W. (1997). An Evening with Berferd In Which a Cracker is Lured, Endured, and Studied. In D. E. Denning (Ed.), *Internet Besieged* (pp. 103–116). New York: ACM Press/Addison-Wesley Publishing Co.

Choi, K., Lee, C. S., & Louderback, E. R. (2020). Historical Evolutions of Cybercrime: From Computer Crime to Cybercrime. In A. M. T. Holt (Ed.), *The Palgrave Handbook of International Cybercrime and Cyberdeviance*. Cham: Palgrave Macmillan. https://doi.org/10.1007/978-3-319-90307-1_2-1.

Chromey, F. R. (2010). *To Measure the Sky: An Introduction to Observational Astronomy*. Cambridge, UK: Cambridge University Press.

Chung, H., Park, J., & Lee, S. (2017). Digital Forensic Approaches for Amazon Alexa Ecosystem. *Digital Investigation 22*, S15–S25. https://doi.org/10.1016/j.diin.2017.06.010.

Ciardhuáin, S. Ó. (2004). An Extended Model of Cybercrime Investigations. *International Journal of Digital Evidence 3* (1), 1–22.

Cisco. (2020). *Cisco Annual Internet Report (2018-2023)*. Retrieved June 23, 2020, from https://www.cisco.com/c/en/us/solutions/collateral/executive-perspectives/annual-internet-report/white-paper-c11-741490.pdf

Clemens, J. (2015). Automatic Classification of Object Code Using Machine Learning. *Digital Investigation the Proceedings of the Fifteenth Annual DFRWS Conference*.

Cohen, W. W. (2015). Enron Email Dataset. Machine Learning Department at Carnegie Mellon University. Retrieved January 18, 2022, from https://www.cs.cmu.edu/~enron/

College of Policing. (2020, March 15). *Investigation Process*. Retrieved from https://www.app.college.police.uk/app-content/investigations/investigation-process/

Cook, R., Evett, I. W., Jackson, G. J., & Lambert, J. (1998). A hierarchy of Propositions: Deciding Which Level to Address in Casework. *Science & Justice 38* (4), 231–239.

Cook, T. (2016). *Blackstone's Senior Investigating Officers' Handbook*, 4th ed. Oxford: Oxford University Press.

Copland v UK, Application no. 62617/00 (European Court of Human Rights (ECtHR) April 3, 2007).

Council of Europe. (2001). *Convention on Cybercrime - European Treaty Series - No. 185*. Budapest.

Curran, J. (2016). The Internet of History: Rethinking the Internet's Past. In J. F. Curran (Ed.), *Misunderstanding the Internet* (pp. 48–84). New York: Routledge.

Custers, B. A. (2021). The Use of Data as Evidence in Dutch Criminal Courts. *European Journal of Crime, Criminal Law and Criminal Justice 29*, 25–46.

Danezis, G., & Sassaman, L. (2003, October). Heartbeat Traffic to Counter (*n*-1) Attacks. *Proceedings of the Workshop on Privacy in the Electronic Society (WPES 2003)*. Washington, DC.

Danezis, G., Dingledine, R., & Mathewson, N. (2003, May). Mixminion: Design of a Type III Anonymous Remailer Protocol. *Proceedings of the 2003 IEEE Symposium on Security and Privacy (S\&P'03)* (pp. 2–15). IEEE Computer Society.

Dean, G. (2000). The Experience of Investigation for Detectives. Doctoral Dissertation, Queensland University of Technology, Brisbane.

Deering, S. E., & Hinden, B. (2017). Internet Protocol, Version 6 (IPv6) Specification. *IETF Request for Comments*. https://doi.org/10.17487/RFC8200

Diaz, C., & Preneel, B. (2004, August). Taxonomy of Mixes and Dummy Traffic. *Proceedings of I-NetSec04: 3rd Working Conference on Privacy and Anonymity in Networked and Distributed Systems* (pp. 215–230). Toulouse.

Dilijonaite, A. (2018). Digital Forensic Readiness. In A. Årnes (Ed.), *Digital Forensics* (pp. 117–145). Hoboken, NJ: John Wiley & Sons.

Dingledine, R. (2013). *Tor and the Silk Road Takedown*. Retrieved from https://blog.torproject.org/tor-and-silk-road-takedown

Dingledine, R., Mathewson, N., & Syverson, P. (2004, August). Tor: The Second-Generation Onion Router. *Proceedings of the 13th USENIX Security Symposium* (pp. 303–320).

Dohr, A., Modre-Opsrian, R., Drobics, M., Hayn, D., Schreier, G., Schreier, G. (2010). The Internet of Things for Ambient Assisted Living. *ITNG2010 - 7th International Conference on Information Technology: New Generations*. 10.1109/ITNG.2010.104

Dorai, G., Houshmand, S., & Baggili, I. (2018). I Know What You Did Last Summer: Your Smart Home Internet of Things and Your iPhone Forensically Ratting You Out. *ACM International Conference Proceeding Series*. https://doi.org/10.1145/3230833.3232814

Dordal, P. L. (2018). The Dark Web. In H. Jahankhani (Ed.), *Cyber Criminology* (pp. 95–117). Cham: Springer.

Douceur, J. (2002, March). The Sybil Attack. In P. Druschel, F. Kaashoek and A. Rowstron (Eds.), *Proceedings of the Peer To Peer Systems: First International Workshop (IPTPS 2002)*, LNCS 2429 (pp. 251–260). Springer-Verlag.

Dror, I. E. (2011). The Paradox of Human Expertise: Why Experts Get It Wrong. In N. Kapur (Ed.), *The Paradoxical Brain* (pp. 177–188). Cambridge: Cambridge University Press.

Dror, I. E. (2014). Practical Solutions to Cognitive and Human Factor Challenges in Forensic Science. *Forensic Science Policy & Management: An International Journal 4* (3–4), 105–113.

Dror, I. E. (2020). Cognitive and Human Factors in Expert Decision Making: Six Fallacies and the Eight Sources of Bias. *Analytical Chemistry 12*, 7998–8004.

Dror, I. E., Morgan, R., Rando, C., & Nakhaeizadeh, S. (2017). The Bias Snowball and the Bias Cascade Effects: Two Distinct Biases that May Impact Forensic Decision Making. *Journal of Forensic Sciences 62* (3), 832–833.

Dror, I. E., Thompson, W. C., Meissner, C. A., Kornfield, I., Krane, D., Saks, M., & Risinger, M. (2015). Letter to the Editor-Context Management Toolbox: A Linear Sequential Unmasking (LSU) Approach for Minimizing Cognitive Bias in Forensic Decision Making. *Journal of Forensic Sciences 60* (4), 1111–1112.

Duerst, M., & Suignard, M. (2005). RFC 3987: Internationalized Resource Identifiers (IRIs). *Request for Comments*. IETF. 10.17487/RFC3987

Eclipse Foundation. (2018). *IoT Developer Surveys*. Retrieved July 16, 2019, from https://iot.eclipse.org/iot-developer-surveys/

ENFSI. (2015a). *ENFSI Guideline for Evaluative Reporting in Forensic Science*. Strengthening the Evaluation of Forensic Results across Europe (STEOFRAE).

ENFSI. (2015b). *Best Practice Manual for the Forensic Examination of Digital Technology, ENFSI-BPM-FOT-01, Version 01 (November 2015)*.

European Council. (2016). *Regulation (EU) 2016/679 of the European Parliament and of the Council of 27 April 2016 on the Protection of Natural Persons with Regard to the Processing of Personal Data and on the Free Movement of Such Data, and Repealing*

Directive 95/46/EC (General Da). Retrieved from http://data.europa.eu/eli/reg/2016/679/2016-05-04

Europol (2016, December 1). *'Avalanche' Network Dismantled in International Cyber Operation.* Den Haag, Netherlands: Europol.

Europol (2019). *Internet Organized Crime Threat Assessment.* The Hague: Europol.

Fahsing, I. A. (2016). *The Making of an Expert Detective. Thinking and Deciding in Criminal Investigations.* Doctoral Dissertation, University of Gothenburg, Gothenburg.

Fan, Z., & de Queiroz, R. (2003, February). Identification of Bitmap Compression History: JPEG Detection and Quantizer Estimation. *IEEE Transactions on Image Processing 12* (2), 230–235.

Flaglien, A. O. (2010). *Cross-Computer Malware Detection in Digital Forensics.* MSc thesis. NTNU, Gjøvik.

Flaglien, A. O. (2018). The Digital Forensics Process. In A. Årnes (Ed.), *Digital Forensics* (pp. 13–49). Hoboken, NJ: Wiley. https://doi.org/10.1002/9781119262442.ch2.

Flaglien, A. O., Mallasvik, A., Mustorp, M., & Årnes, A. (2011, November 1). Storage and Exchange Formats for Digital Evidence. *Digital Investigation 8*, 122–128.

Frank, J., Eisenhofer, T., Schönherr, L., Fischer, A., Kolossa, D., & Holz, T. (2020). Leveraging Frequency Analysis for Deep Fake Recognition. *Proceedings of the 37th International Conference on Machine Learning, Volume 119 of Proceedings of Marchine Learning Research* (pp. 3247–3258). PMLR.

Franke, K., & Srihari, S. N. (2008). Computational Forensics: An Overview. In S. N. Srihari and K. Franke (Eds.), *Computational Forensics.* Springer. Retrieved from https://www.researchgate.net/publication/221577624_Computational_Forensics_An_Overview.

Gallagher, A. (2005). Detection of Linear and Cubic Interpolation in JPEG Compressed Images. *The 2nd Canadian Conference on Computer and Robot Vision (CRV '05), Number 2* (pp. 65–72). IEEE.

Garg, D., Goel, P., Kansaswamy, G., Ganatra, A., & Kotecha, K. (2019). A Roadmap to Deep Learning: A State-of-the-Art Step Towards Machine Learning. In *Communications in Computer and Information Science*, vol. 955 (pp. 160–170). Singapore: Springer.

Garrett, B., & Mitchell, G. (2016). Forensics and Fallibility: Comparing the Views of Lawyers and Judges. *West Virginia Law Review 119*, 2016–2063.

Gault, M. (2015, December 20). *The CIA Secret to Cybersecurity That No One Seems to Get.* wired.com, Retrieved from https://www.wired.com/2015/12/the-cia-secret-to-cybersecurity-that-no-one-seems-to-get/

Goel, S., Robson, M., Polte, M., & Sirer, E. G. (2003, February). *Herbivore: A Scalable and Efficient Protocol for Anonymous Communication.* Technical Report. Ithaca, NY: Cornell University.

Goldberg, I. (2000, December). *A Pseudonymous Communications Infrastructure for the Internet.* PhD dissertation, UC Berkeley.

Goldberg, I. (2002, April). Privacy-Enhancing Technologies for the Internet, II: Five years Later. In R. Dingledine and P. Syverson (Eds.), *Proceedings of Privacy Enhancing Technologies Workshop (PET 2002)*, LNCS 2482 (pp. 1–12). Springer-Verlag. Retrieved from https://dl.acm.org/doi/10.5555/1765299.1765300.

Goldberg, I., Wagner, D., & Brewer, E. (1997, February). Privacy-Enhancing Technologies for the Internet. *Proceedings of the 42nd IEEE Spring COMPCON* (pp. 103–109). IEEE Computer Society Press.

Goldschlag, D. M., Reed, M. G., & Syverson, P. F. (1996, May). Hiding Routing Information. In R. Anderson (Ed.), *Proceedings of Information Hiding: First International Workshop*, LNCS 1174 (pp. 137–150). Springer-Verlag. Retrieved from https://link.springer.com/chapter/10.1007/3-540-61996-8_37.

Goodfellow, I., Puget-Abadie, J., Mirza, M., Xu, B., Warde-Farley, D., Ozair, S., . . . Bengio, Y. (2014). Generative Adversarial Nets. In *Advances in Neural Information Processing Systems*, vol. 27. Curran Associates, Inc. Retrieved from https://scholar.google.com/scholar_lookup?arxiv_id=1406.2661.

Goodin, D. (2017). *BrickerBot, the Permanent Denial-of-Service Botnet, is Back with a Vengeance.* ArsTechnica, Retrieved June 26, 2019, from https://arstechnica.com/information-technology/2017/04/brickerbot-the-permanent-denial-of-service-botnet-is-back-with-a-vengeance/

Grajeda, C., Breitinger, F., & Baggili, I. (2017). Availability of Datasets for Digital Forensics – and What is Missing. *Digital Investigation 22*, 94–105.

Granhag, P. A., & Ask, K. (2021). Psykologiska Perspektiv på Bevisvärdering. In P. A. Granhag, L. A. Strömvall, K. Ask and S. Landstrøm (Eds.), *Handbok i Rättspsykologi*. Stockholm: Liber.

Greenberg, A. (2019, December 18). *Meet the Mad Scientist Who Wrote the Book on How toHuntHackers.*Wired.com,Retrievedfromhttps://www.wired.com/story/meet-the-mad-scientist-who-wrote-the-book-on-how-to-hunt-hackers/

Greenstein, S. (2015). *How the Internet Became Commercial: Innovation, Privatization, and the Birth of a New Network*. Princeton, NJ: Princeton University Press.

Greenwald, G. (2014). *No Place to Hide: Edward Snowden, the NSA, and the U.S. Surveillance State*. London: Metropolitan Books.

Guinard, D., Trifa, V., Karnouskos, S., Spiess, P., & Savio, D. (2010). Interacting with the SOA-Based Internet of Things: Discovery, Query, Selection, and On-Demand Provisioning of Web Services. *IEEE Transactions on Services Computing 3*, 223–235. https://doi.org/10.1109/TSC.2010.3.

Gülcü, C., & Tsudik, G. (1996, February). Mixing E-mail With Babel. *SNDSS '96: Proceedings of the 1996 Symposium on Network and Distributed System Security (SNDSS '96)* (pp. 2–16). Washington, DC: IEEE Computer Society.

Hald, C., & Rønn, K. V. (2013). Inledning. In C. R. Hald (Ed.), *Om at Opdage* (pp. 15–54). Frederiksberg C: Samfundslitteratur.

Halford v UK, Application no. 20605/92 (European Court of Human Rights (ECtHR) June 25, 1997).

Hamm, J. (2018). Computer Forensics. In A. Årnes (Ed.), *Digital Forensics*. Hoboken, NJ: John Wiley & Sons, Ltd.

Harris, D. O. (2018). *Law of the European Convention on Human Rights*. Oxford: Oxford University Press.

He, W., Yan, G., & Xu, L. D. (2014). Developing Vehicular Data Cloud Services in the IoT Environment. *IEEE Transactions on Industrial Informatics 10* (2), 1587–1595. https://doi.org/10.1109/TII.2014.2299233.

Helmers, S. (1997, September). *A Brief History of anon.penet.fi - The Legendary Anonymous Remailer*. CMC Magazine. Retrieved from https://www.december.com/cmc/mag/1997/sep/helmers.html

Helstrup, T., & Kaufmann, G. (2000). *Kognitiv Psykologi*. Bergen: Fagbokforlaget.

Heuer, R. (1999). *Psychology of Intelligence Analysis*. Washington, DC: Center for the Study of Intelligence, Central Intelligence Agency.

Hintz, A. (2002, April). Fingerprinting Websites Using Traffic Analysis. In R. Dingledine and P. Syverson (Eds.), *Proceedings of Privacy Enhancing Technologies Workshop (PET 2002)*, LNCS 2482 (pp. 171–178). Springer-Verlag. Retrieved from https://link.springer.com/chapter/10.1007/3-540-36467-6_13.

Holt, T. J., Bossler, A. M., & Seigfried-Spellar, K. (2018). *Cybercrime and Digital Forensics. An Introduction*. London: Routledge.

Horsman, G., & Sunde, N. (2020). Part 1: The Need for Peer Review in Digital Forensics. *Forensic Science International: Digital Investigation 35*, 301062.

Hotelling, H. (1935, December). Relations Between Two Sets of Variates. *Biometrika 28* (3/4), 321.

Huang, H., Guo, W., & Zhang, Y. (2018, December). Detection of Copy-Move Forgery in Digital Images Using SIFT Algorithm. *IEEE Pacific-Asia Workshop on Computational Intelligence and Industrial Application, volume 2* (pp. 272–276).

Huang, Y., Guo, Q., Li, J., Juefei-Xu, F., Ma, L., Miao, W., Wang, R., Xie, X., Liu, Y., & Pu, G. (2020). FakePolisher; Making DeepFakes More Detection-Evasive by Shallow Reconstruction. arXiv.

Hung, M. (2017). *Leading the IoT, Gartner Insights on How to Lead in a Connected World*. Gartner Research. Retrieved from https://www.gartner.com/imagesrv/books/iot/iotEbook_digital.pdf.

Hunton, P. (2009). The Growing Phenomenon of Crime and the Internet: A Cybercrime Execution and Analysis Model. *Computer Law & Security Review 25* (6), 528–535.

Hunton, P. (2011a). A Rigorous Approach to Formalising the Technical Investigation Stages of Cybercrime and Criminality within a UK Law Enforcement Environment. *Digital Investigation 7* (3–4), 105–113.

Hunton, P. (2011b). The Stages of Cybercrime Investigations: Bridging the Gap between Technology Examination and Law Enforcement Investigation. *Computer Law and Security Review 27* (1), 61–67. https://doi.org/10.1016/j.clsr.2010.11.001.

Hutchins, E. M. (2011). Intelligence-Driven Computer Network Defense Informed by Analysis of Adversary Campaigns and Intrusion Kill Chains. *Leading Issues in Information Warfare & Security Research 1*, 80–91.

IANA. (2018). *IANA IPv6 Special-Purpose Address Registry*. Retrieved November 20, 2018, from https://www.iana.org/assignments/iana-ipv6-special-registry/iana-ipv6-special-registry.xhtml

IAPR. (2022, January 18). *IAPR - International Association of Pattern Recognition*. Retrieved from https://iapr.org/

IEEE (2022). IEEE Standard for Prefixes for Binary Multiples. IEEE Std 1541-2021. https://doi.org/10.1109/IEEESTD.2022.9714443

Innes, M. (2003). *Investigating Murder: Detective Work and the Police Response to Criminal Homicide*. Oxford: Oxford University Press.

Innes, M. (2007). Investigation Order and Major Crime Inquiries. In T. W. Tim Newburn (Ed.), *Handbook of Criminal Investigation* (pp. 255–276). New York: Willian Publishing.

Intel. (2018). *A Guide to the Internet of Things Infographic*. Retrieved August 26, 2019, from https://www.intel.com/content/www/us/en/internet-of-things/infographics/guide-to-iot.html

Internet World Stats. (2021, December 27). *World Internet Users and 2021 Population Stats.* Retrieved from https://www.internetworldstats.com/stats.htm

Interpol. (2018, August 28). *Cybercrime.* Interpol. Retrieved from https://www.interpol.int/Crime-areas/Cybercrime/Cybercrime

ISACA. (2022, January 18). *Information Systems Audit Control Associations (ISACA).* Retrieved from https://www.isaca.org

ISO/IEC. (2012). *ISO/IEC 27037:2012 Information Technology -- Security Techniques -- Guidelines for Identification, Collection, Acquisition and Preservation of Digital Evidence.*

Jackson, G. (2009). Understanding Forensic Science Opinions. In J. Fraser and R. Williams (Eds.), *Handbook of Forensic Science* (pp. 419–445). Portland: Willian.

Jackson, G. (2011). *The Development of Case Assessment and Interpretation (CAI) in Forensic Science.* PhD thesis, University of Abertay Dundee.

Jackson, G., Aitken, C., & Roberts, P. (2015). Case Assessment and Interpretation of Expert Evidence. Guidance for Judges, Lawyers, Forensic Scientists and Expert Witnesses. Practitioner Guide No. 4. Royal Statistic Society, https://www.researchgate.net/publication/273698920_Case_Assessment_and_Interpretation_of_Expert_Evidence_Guidance_for_Judges_Lawyers_Forensic_Scientists_and_Expert_Witnesses.

Jammes, F., & Smit, H. (2005). Service-Oriented Paradigms in Industrial Automation. *IEEE Transactions on Industrial Informatics 1*, 62–70. https://doi.org/10.1109/TII.2005.844419.

Janis, I. L. (1972). *Victims of Groupthink: A Psychological Study of Foreign-Policy Decisions and Fiascoes.* Boston, MA: Houghton Mifflin.

Jara, A. J., Ladid, L., & Skarmeta, A. (2013). The Internet of Everything through IPv6: An Analysis of Challenges, Solutions and Opportunities. *Journal of Wireless Mobile Networks, Ubiquitous Computing, and Dependable Applications 4* (3), 97–118.

Jiang, X., Xu, Q., Sun, T., Li, B., & He, P. (2020). Detection of HEVC Double Compression with the Same Coding Parameters Based on Analysis of Intra Coding Quality Degradation Process. *IEEE Transactions on Information Forensics and Security 15*, 250–263.

Johnson, B., Caban, D., Krotofil, M., Scali, D., Brubaker, N., & Glyer, C. (2017). *Attackers Deploy New ICS Attack Framework "TRITON" and Cause Operational Disruption to Critical Infrastructure.* FireEye Threat Research. Retrieved December 17, 2018, from https://www.fireeye.com/blog/threat-research/2017/12/attackers-deploy-new-ics-attack-framework-triton.html

Johnson, M. K., & Farid, H. (2005). Exposing Digital Forgeries by Detecting Inconsistencies in Lighting. *Proceedings of the 7th Workshop on Multimedia and Security MM&Sec'05, Number 1* (pp. 1–10). New York: ACM Press.

Jones, D., Grieve, J., & Milne, B. (2008). The Case to Review Murder Investigations. *Policing: A Journal of Policy and Practice 2* (4), 470–480.

Josephson, J. R., & Josephson, S. G. (1996). *Abductive Inference: Computation, Philosophy, Technology.* Cambridge: Cambridge University Press.

Juniper Research. (2020, March 31). *IoT Connections to Reach 83 Billion by 2024, Driven by Maturing Industrial Use Cases.* Retrieved from https://www.juniperresearch.com/press/iot-connections-to-reach-83-bn-by-2024

Kaaniche, N., Kaaniche, N., Laurent, M., Laurent, M., & Belguith, S. (2020). Privacy enhancing technologies for solving the privacy-personalization paradox: Taxonomy and survey. *Journal of Network and Computer Applications 171*, 102807.

Karras, T., Laine, S., Aittala, M., Hellsten, J., Lehtinen, J., & Aila, T. (2020). Analyzing and Improving the Image Quality of StyleGAN. *IEEE/CVF Conference on Computer Vision and Pattern Recognition (CVPR)* (pp. 8107–8116). IEEE.

Katasonov, A., Kaykova, O., Khriyenko, O., Nikitin, S., & Terziyan, V. (2008). Smart Semantic Middleware for the Internet of Things. *ICINCO 2008 - Proceedings of the 5th International Conference on Informatics in Control, Automation and Robotics* (pp. 169–178).

Kersholt, J., & Eikelboom, A. (2007). Effects of Prior Interpretation on Situation Assessment in Crime Analysis. *Journal of Behavioral Decision Making 20* (5), 455–465.

Kjølbro, J. F. (2017). *Den Europæiske Menneskerettighedskonvention for Praktikere*, 4th ed. København: Jurist- og Økonomforbundet. Retrieved from https://www.worldcat.org/title/europiske-menneskerettighedskonvention-for-praktikere/oclc/971130889?referer=di&ht=edition.

Klass and Others v Germany, Application no. 5029/71 (European Court of Human Rights (ECtHR) September 6, 1978).

Kolias, C., Kambourakis, G., Stavrou, A., & Voas, J. (2017). DDoS in the IoT: Mirai and Other Botnets. *Computer 50* (7), 80–84.

Kozhuharova, D. A. (2021). *From Mobile Phones to Court – A Complete FORensic Investigation Chain Targeting MOBILE Devices (FORMOBILE) D2.2 Criminal Procedure Report.* European Union's Horizon 2020 - Research and Innovation Framework Programme.

Krebs, B. (2017, January 18). *Who is Anna-Senpai, the Mirai Worm Author?* Krebs on Security.RetrievedMay19,2018,fromhttps://krebsonsecurity.com/2017/01/who-is-anna-senpai-the-mirai-worm-author/

Langner, R. (2011). Stuxnet: Dissecting a Cyberwarfare Weapon. *IEEE Security and Privacy 9* (3), 49–51.

Larson, S. (2017). *A Smart Fish Tank Left a Casino Vulnerable to Hackers.* CNN Business. Retrieved December 18, 2018, from https://money.cnn.com/2017/07/19/technology/fish-tank-hack-darktrace/index.html

Lee, E. A. (2008). Cyber Physical Systems: Design Challenges. *Proceedings - 11th IEEE Symposium on Object/Component/Service-Oriented RealTime Distributed Computing, ISORC 2008* (pp. 363–369).

Le-Khac, N. A., Jacobs, D., Nijhoff, J., Bertens, K., & Choo, K. K. (2018). Smart Vehicle Forensics: Challenges and Case Study. *Future Generation Computer Systems 109*, 500–510. https://doi.org/10.1016/j.future.2018.05.081.

Lentz, L. W. (2019). *Politiets Hemmelige Efterforskning på Internet.* Aalborg: Aalborg University.

Levine, B. N., Reiter, M. K., Wang, C., & Wright, M. (2004). Timing Attacks in Low-Latency Mix-Based Systems (Extended Abstract). In A. Juels (Ed.), *Financial Cryptography, 8th International Conference, FC 2004*, LNCS 3110 (pp. 251–265). Springer-Verlag. Retrieved from https://link.springer.com/chapter/10.1007/978-3-540-27809-2_25.

Ligh, M. H., Case, A., Levy, J., & Walters, A. (2014). *The art of memory forensics: detecting malware and threats in windows, linux, and Mac memory,* John Wiley & Sons, Inc.

Lim, S. R. (2015). Identifying Management Factors for Digital Incident Responses on Machine-to-Machine Services. *Digital Investigation 14*, 46–52.

Lin, S., & Zhang, L. (2005). Determining the Radiometric Response Function from a Single Grayscale Image. *IEEE Computer Society Conference on Computer Vision and Pattern Recognition (CVPR'05)*, volume 2 (pp. 66–73).

Lin, W., Tjoa, S., & Liu, K. (2009, September). Digital Image Source Coder Forensics via Intrinsic Fingerprints. *IEEE Transactions on Information Forensics and Security 4* (3), 460–475.

Lindsay, P., & Norman, D. (1977). *Human Information Processing: An Introduction to Psychology*. New York: Academic Press.

MacDermott, Á., Baker, T., & Shi, Q. (2018). IoT Forensics: Challenges for the IoA Era. *2018 9th IFIP International Conference on New Technologies, Mobility and Security (NTMS)*.

Maguire, M. (2003). Criminal Investigation and Crime Control. In T. Newburn (Ed.), *Handbook of Policing* (pp. 363–393). Cullompton: Willian.

Maher, R. C. (2015). Lending an Ear in the Courtroom: Forensic Acoustics. *Acoustics Today 11* (3), 22–29.

Maher, R. C. (2018). *Principles of Forensic Audio Analysis, Modern Acoustics and Signal Processing*. Springer International Publishing. Retrieved from https://www.amazon.com/Principles-Forensic-Analysis-Acoustics-Processing/dp/3319994522.

Marriot Webster. (2022, January 23). *Marriot Webster Dictionary*. Retrieved from https://www.merriam-webster.com/dictionary/cybersecurity

McKemmish, R. (2008). When is Digital Evidence Forensically Sound? In *IFIP International Conference on Digital Forensics* (pp. 3–15). Boston, MA: Springer.

Meffert, C., Clark, D., Baggili, I., & Breitinger, F. (2017). Forensic State Acquisition from Internet of Things (FSAIoT). *Proceedings of the 12th International Conference on Availability, Reliability and Security - ARES '17*. https://doi.org/10.1145/3098954.3104053

Meiklejohn, S., Pomarole, M., Jordan, G., Levchenko, K., Damon, M., Voelker, G. M., & Savage, S. (2013, October). A Fistful of Bitcoins: Characterizing Payments Among Men with No Names. *IMC '13: Proceedings of the 2013 Conference on Internet Measurement* (pp. 127–140).

Michael, K., & Hayes, A. (2016). High-Tech Child's Play in the Cloud: Be Safe and Aware of the Difference between Virtual and Real. *IEEE Consumer Electronics Magazine 5* (1), 123–128. https://doi.org/10.1109/MCE.2015.2484878.

Milani, S., Fontani, M., Bestagini, P., Piva, A., Tagliasacchi, M., & Tubaro, S. (2012a, August). An Overview on Video Forensics. In *APSIPA Transactions on Signal and Information Processing*, vol. 1. Retrieved from https://ieeexplore.ieee.org/abstract/document/6288362.

Milani, S., Tagliasacchi, M., & Tubaro, S. (2012b). Discriminating Multiple JPEG Compression Using First Digit Features. *IEEE International Conference on Accoustics, Speech and Signal Processing - Proceedings* (pp. 2253–2256). ICASSP.

Miller, C., & Valasek, C. (2015). *Remote Exploitation of an Unaltered Passenger Vehicle*. Defcon 23. Retrieved from http://illmatics.com/Remote Car Hacking.pdf

MITRE. (2022, January 23). *MITRE ATT&CK™*. Retrieved from attack.mitre.org

Möller, U., Cottrell, L., Palfrader, P., & Sassaman, L. (2004, December). Mixmaster Protocol Version 2.

Montenegro, G., Kushalnagar, N., Hui, J. W., & Culler, D. E. (2007). *RFC 4944: Transmission of IPv6 Packets over IEEE 802.15.4 Networks*. Networking Group, IETF.

Moreno, M. V., Skarmeta, A. F., & Jara, A. J. (2015). How to Intelligently Make Sense of Real Data of Smart Cities. *2015 International Conference on Recent Advances in Internet of Things, RIoT 2015*. https://doi.org/10.1109/RIOT.2015.7104899

Mqtt.org. (n.d.). *MQTT*. Retrieved July 16, 2019, from https://mqtt.org

Muir, R. (2021). *Taking Prevention Seriously: The Case for a Crime and Harm Prevention System. Insight Paper 3*. The Police Foundation.

Muniswamy-Reddy, K.-K., Holland, D. A., Braun, U., & Seltzer, M. (2006). Provenance-aware Storage Systems. *Proceedings of the USENIX Annual Technical Conference* (pp. 43–56).

Mützel, D. (2017, November 8). *Meet the Hacker Who Busts Child Pornographers on the Dark Net*. Motherboard Tech by VICE. Retrieved from https://motherboard.vice.com/en_us/article/ywbmyb/meet-the-hacker-who-busts-child-pornographers-on-the-dark-net

N. Harris Computer Corporation. (2022, January 18). *i2 Intelligence Analysis and Advanced Analytics Platform*. Retrieved from https://i2group.com

Nakamoto, S. (2009, May 24). Bitcoin: A Peer-to-Peer Electronic Cash System, https://www.researchgate.net/publication/228640975_Bitcoin_A_Peer-to-Peer_Electronic_Cash_System.

National Institute of Standards and Technology (NIST) (2006). *Special Publication 800-86; Guide to Integrating Forensic Techniques into Incident Response*. Gaithersburg, MD: National Institute of Standards and Technology.

Ng, T.-T., Chang, S.-F., & Tsui, M.-P. (2007). Using Geometry Invariants for Camera Response Function Estimation. *IEEE Conference on Computer Vision and Pattern Recognition* (pp. 1–8). IEEE.

Nickerson, R. S. (1998). Confirmation Bias: A Ubiquitous Phenomenon in Many Guises. *Review of General Psychology 2* (2), 175.

Norman-dommen, Straffeloven §145 og §393: Påtalemyndigheten mot X Systems AS og A (Høyesterett (Norway) December 15, 1998).

Odom, N. R., Lindmar, J. M., Hirt, J., & Brunty, J. (2019). Forensic Inspection of Sensitive User Data and Artifacts from Smartwatch Wearable Devices. *Journal of Forensic Sciences 64* (6), 1673–1686. https://doi.org/10.1111/1556-4029.14109.

Oh, H., Rizo, C., Enkin, M., & Jadad, A. (2005). What is eHealth (3): A Systematic Review of Published Definitions. *Journal of Medical Internet Research 7* (1), e1. https://doi.org/10.2196/jmir.7.1.e1.

Olsson v Sweden (No 1) (plenary), Application no. 10465/83 (European Court of Human Rights (ECtHR) March 24, 1998).

Omand, D. (2021). The Historical Backdrop. In S. Stenslie, L. Haugom and B. H. Vaage (Eds.), *Intelligence Analysis in the Digital Age* (pp. 12–25). London: Routledge.

Oriwoh, E., & Sant, P. (2013). The Forensics Edge Management System: A Concept and Design. *2013 IEEE 10th International Conference on Ubiquitous Intelligence and Computing and 2013 IEEE 10th International Conference on Autonomic and Trusted Computing* (pp. 544–550).

Oriwoh, E., Jazani, D., Epiphaniou, G., & Sant, P. (2013). Internet of Things Forensics: Challenges and Approaches. *Proceedings of the 9th IEEE International Conference on Collaborative Computing: Networking, Applications and Worksharing* (pp. 608–615).

Page, H., Horsman, G., Sarna, A., & Foster, J. (2018). A Review of Quality Procedures in the UK Forensic Sciences: What Can the Field of Digital Forensics Learn? *Science & Justice 59* (1), 83–92. https://doi.org/10.1016/j.scijus.2018.09.006.

Parker, D. B. (1976). *Crime by Computer*. New York: Charles Scribner's Sons.

Parra, L., Sendra, S., Garcia, L., & Lloret, J. (2018). Design and Deployment of Low-Cost Sensors for Monitoring the Water Quality and Fish Behavior in Aquaculture Tanks During the Feeding Process. *Sensors (Switzerland) 18* (3), 750. https://doi.org/10.3390/s18030750.

Paul, K. (2015, February 8). *Russia Wants to Block Tor, But It Probably Can't.* Motherboard. Retrieved from https://motherboard.vice.com/en_us/article/ypwevy/russia-wants-to-block-tor-but-it-probably-cant

Pearson, K. (1901). On Lines and Planes of Closest Fit to Systems of Points in Space. *The London, Edinburgh, and Dublin Philosophical Magazine and Journal of Science 2* (11), 559–572.

Pfitzmann, A., & Hansen, M. (2007, July). Anonymity, Unlinkability, Undetectability, Unobservability, Pseudonymity, and Identity Management - A Consolidated Proposal for Terminology.

Piva, A. (2013). An Overview on Image Forensics. *ISRN Signal Processing 2013*, 1–22.

Politiet. (2019). *Cybercrime*. Retrieved October 30, 2019, from https://www.politiet.no/en/rad/cybercrime/

Pollitt, M., Casey, E., Jaquet-Chiffelle, D., & Gladyshev, P. (2018). *A Framework for Harmonizing Forensic Science Practices and Digital/Multimedia Evidence*. USA: The Organization of Scientific Area Committees for Forensic Science (OSAC). Retrieved from https://www.nist.gov/osac/framework-harmonizing-forensic-science-practices-and-digitalmultimedia-evidence.

Popescu, A. C., & Farid, H. (2005). Exposing Digital Forgeries in Color Filter Array Interpolated Images. *IEEE Transactions on Signal Processing 53* (10 II), 3948–3959.

Popov, S. (2018). *The Tangle*. iota.org. Retrieved from https://iota.org

Popper, K. (1963). *Conjectures and Falsifications*. London: Routledge.

Quick, D., & Choo, K. K. (2016). Big Forensic Data Reduction: Digital Forensic Images and Electronic Evidence. *Cluster Computing 19* (2), 723–740.

Quick, D., & Choo, K.-K. (2018). IoT Device Forensics and Data Reduction. *IEEE Access 6*, 47566–47574.

Ramaswamy, L., Liu, L., & Iyengar, A. (2007). Scalable Delivery of Dynamic Content Using a Cooperative Edge Cache Grid. *IEEE Transactions on Knowledge and Data Engineering 19* (5), 614–630. https://doi.org/10.1109/TKDE.2007.1031.

Rathod, D., & Chowdhary, G. (2018). Survey of Middlewares for Internet of Things. *Proceedings of the 2018 International Conference on Recent Trends in Advanced Computing, ICRTAC-CPS 2018*. https://doi.org/10.1109/ICRTAC.2018.8679249

Reed, M. G., Syverson, P. F., & Goldschlag, D. M. (1996, December). Proxies for Anonymous Routing. *ACSAC '96: Proceedings of the 12th Annual Computer Security Applications Conference* (pp. 95–104). Washington, DC: IEEE Computer Society.

Reiter, M., & Rubin, A. (1998, June). Crowds: Anonymity for Web Transactions. *ACM Transactions on Information and System Security (TISSEC) 1*, 66–92.

Reyna, A., Martín, C., Chen, J., Soler, E., & Díaz, M. (2018). On Blockchain and Its Integration with IoT. Challenges and Opportunities. *Future Generation Computer Systems 88*, 173–190. https://doi.org/10.1016/j.future.2018.05.046.

Rønn, K. V. (2022). Intelligence Analysis in the Digital Age. In S. A. Stenslie (Ed.), *Intelligence Analysis in the Digital Age*. London: Routledge Taylor & Francis Group.

Rosen, R., Von Wichert, G., Lo, G., & Bettenhausen, K. D. (2015). About the Importance of Autonomy and Digital Twins for the Future of Manufacturing. *IFAC-PapersOnLine 28* (3). https://doi.org/10.1016/j.ifacol.2015.06.141.

Ruan, K., Carthy, J., Kechadi, T., & Crosbie, M. (2011). Cloud Forensics. In G. Peterson and S. Shenoi (Eds.), *Advances in Digital Forensics VII*. Berlin, Heidelberg: Springer Berlin Heidelberg.

Salman, O., Elhajj, I., Kayssi, A., & Chehab, A. (2015). Edge Computing Enabling the Internet of Things. *2015 IEEE 2nd World Forum on Internet of Things (WF-IoT)*. https://doi.org/10.1109/WF-IoT.2015.7389122

Sanchez, L., Muñoz, L., Galache, J., Sotres, P., Santana, J. R., Gutierrez, V., . . . Pfisterer, D. (2014). SmartSantander: IoT Experimentation Over a Smart City Testbed. *Computer Networks 61*, 217–238.

Sandvik, J.-P., & Årnes, A. (2018). The Reliability of Clocks as Digital Evidence under Low Voltage Conditions. *Digital Investigation 24*, S10–S17.

Satakunnan Markkinapörssi Oy and Satamedia Oy v Finland, Judgment (GC), Application no. 931/13 (European Court of Human Rights (ECtHR) June 27, 2017).

Schermer, B., Georgieva, I., van der Hof, S., & Koops, B.-J. (2019). Legal Aspects of Sweetie 2.0. In I. B. Schermer, I. Georgieva, S. van der Hof and B.-J. Koops (Eds.), *Sweetie 2.0 - Using Artificial Intelligence to Fight Webcam Child Sex Tourism* (pp. 1–93). The Hague: Springer Information Technology and Law Series. Retrieved from https://research.tilburguniversity.edu/en/publications/legal-aspects-of-sweetie-20-2.

Schmitt, M. N. (2017). *Tallinn Manual 2.0 on the International Law Applicable to Cyber Operations*. Cambridge: Cambridge University Press.

Schwab, K. (2016). *The Fourth Industrial Revolution*. New York: Crown Business.

Serjantov, A., Dingledine, R., & Syverson, P. (2002, October). From a Trickle to a Flood: Active Attacks on Several Mix Types. In F. Petitcolas (Ed.), *Proceedings of Information Hiding Workshop (IH 2002)*, LNCS 2578 (pp. 36–52). Springer-Verlag. Retrieved from https://link.springer.com/chapter/10.1007/3-540-36415-3_3?noAccess=true.

Shah, A., Rajdev, P., & Kotak, J. (2019). *Memory Forensic Analysis of MQTT Devices*. Retrieved from https://www.researchgate.net/publication/335318992_Memory_Forensic_Analysis_of_MQTT_Devices.

Sherwood, R., Bhattacharjee, B., & Srinivasan, A. (2002, May). P5: A Protocol for Scalable Anonymous Communication. *Proceedings of the 2002 IEEE Symposium on Security and Privacy (S\&P'02)* (pp. 58–70). IEEE Computer Society.

Silva, B. M., Rodrigues, J. J. P. C., de la Torre Díez, I., Lopez-Coronado, M., & Saleem, K. (2015, August). Mobile-health: A Review of Current State in 2015. *Journal of biomedical informatics 56*, 265–272. https://doi.org/10.1016/J.JBI.2015.06.003.

Skjold, J. S. (2019). Suverenitet, Jurisdiksjon og Beslag i Informasjon på Server i Utlandet. *Lov og Rett 58*, 617–639.

Smit, N., Morgan, R., & Lagnado, D. (2018). A Systematic Analysis of Misleading Evidence in Unsafe Rulings in England and Wales. *Science & Justice 58* (2), 128–137.

Societeé Colas Est and Others v France, Application no. 37971/97 (European Court of Human Rights (ECtHR) April 16, 2002).

Soos, G. K. (2018). IoT Device Lifecycle – A Generic Model and a Use Case for Cellular Mobile Networks. *2018 IEEE 6th International Conference on Future Internet of Things and Cloud (FiCloud)* (pp. 176–183).

Spiess, P., Karnouskos, S., Guinard, D., Savio, D., Baecker, O., Souza, L. M., & Trifa, V. (2009). SOA-based Integration of the Internet of Things in Enterprise Services. *2009 IEEE International Conference on Web Services, ICWS 2009*. https://doi.org/10.1109/ICWS.2009.98

Stelfox, P. (2013). *Criminal Investigation: An Introduction to Principles and Practice.* London: Routeledge.

Stenslie, S. A. (2021). An Old Activity in a New Age. In S. A. Stenslie (Ed.), *Intelligence Analysis in the Digital Age.* London: Routledge Taylor & Francis Group.

Stevens, W. R. (1999). *UNIX Network Programming,* (2). ed., vol. 2. Upper Saddle River, NJ: Prentice Hall.

Stoll, C. (1989). *The Cuckoo's Egg: Tracking a Spy Through the Maze of Computer Espionage,* 1st ed. London: The Bodley Head Ltd.

Stumer, A. (2010). *The Presumption of Innocence: Evidential and Human Rights.* Oxford and Portland: Hart Publishing.

Sunde, I. M. (2018). Cybercrime Law. In A. Årnes and A. Årnes (Eds.), *Digital Forensics.* Hoboken, NJ: John Wiley & Sons, Ltd.

Sunde, N. (2021a). What Does a Digital Forensic Opinion Look Like? A Comparative Study of Digital Forensics and Forensic Science Reporting Practices. *Science & Justice 5,* 586–596.

Sunde, I. M. (2021a). *Effektiv, Tillitvekkende og Rettssikker Behandling av Databevis.* Oslo: Justis- og beredskapsdepartementet.

Sunde, I. M. (2021b). Norske Verdioppfatninger og Kriminelles Forventninger Som Skranker for Bruk av Bevis Fremskaffet av Utenlandsk Politi. *Tidsskrift for strafferett 21,* 207–210.

Sunde, N. (2017). *Non-technical Sources of Errors When Handling Digital Evidence within a Criminal Investigation.* Master's thesis, Faculty of Technology and Electrical Engineering, Norwegian University of Science and Technology. Gjøvik, Norwegian University of Science and Technology.

Sunde, N. (2020). Structured Hypothesis Development in Criminal Investigation. *The Police Journal,* 1–17. Retrieved from https://journals.sagepub.com/doi/abs/10.1177/003 2258X20982328#:~:text=The%20Structured%20Hypothesis%20Development%20 in,a%20broad%20and%20objective%20investigation.

Sunde, N., & Dror, I. (2019). Cognitive and Human Factors in Digital Forensics: Problems, Challenges and the Way Forward. *Digital Investigation 29,* 101–108. https://doi. org/10.1016/j.diin.2019.03.011.

Sunde, N., & Dror, I. E. (2021). A Hierarchy of Expert Performance (HEP) Applied to Digital Forensics: Reliability and Biasability in Digital Forensics Decision Making. *Forensic Science International: Digital Investigation 37,* 301175.

Sunde, N., & Horsman, G. (2021). Part 2: The Phase Oriented Advice and Review Structure (PARS) for Digital Forensic Investigations. *Forensic Science International: Digital Investigation 36,* 301074.

Sunde, N., & Sunde, I. M. (2019). *Det Digitale er et Hurtigtog!: Vitenskapelige Perspektiver på Politiarbeid, Digitalisering og Teknologi. Nordic Journal of Studies in Policing 8,* 1–21.

Sunde, N., & Sunde, I. M. (2021). Conceptualizing an AI-based Police Robot for Preventing Online Child Sexual Exploitation and Abuse: Part I – The Theoretical and Technical Foundations for PrevBOT. *Nordic Journal of Studies in Policing 8* (2), 1–21.

Sunde, N., & Sunde, I. M. (2022) Accepted for publication. Conceptualizing an AI-based Police Robot for Preventing Online Child Sexual Exploitation and Abuse. Part II – Legal Analysis of PrevBOT.

SWGDE. (2018). *Establishing Confidence in Digital Forensic Results by Error Mitigation Analysis Version 2.0 (November 20, 2018).* SWGDE.

Tao, F., Zhang, L., Venkatesh, V. C., Luo, Y., & Cheng, Y. (2011). Cloud Manufacturing: A Computing and Service-Oriented Manufacturing Model. *Proceedings of the Institution of Mechanical Engineers, Part B: Journal of Engineering Manufacture 225* (10), 1969–1976.

Thompson, W. C., & Schumann, E. I. (1987). Interpretation of Statistical Evidence in Criminal Trials - The Prosecutor's Fallacy and the Defence Attorney's Fallacy. *Law and Human Behavior 3*, 167–187.

Tilstone, W. J., Hastrup, M. L., & Hald, C. (2013). *Fischer's Techniques of Crime Scene Investigation*. Boca Raton, FL: CRC Press, Taylor and Francis Group, LLC.

Tou, J. Y., Yee Lau, P., & Park, S.-K. (2014). Video Forensics on Block Detection with Deblocking Filter in H.264/AVC. *Signal and Information Processing Association Annual Summit and Conference (APSIPA)* (pp. 1–9). IEEE.

Trivedi, K. S., Grottke, M., & Andrade, E. (2010). Software Fault Mitigation and Availability Assurance Techniques. *International Journal of Systems Assurance Engineering and Management 1* (4), 340–350. https://doi.org/10.1007/s13198-011-0038-9.

Tversky, A., & Kahneman, D. (1974). Judgment under Uncertainty: Heuristics and Biases. *Science 185* (4157), 1124–1131.

Valenzise, G., Tagliasacchi, M., & Tubaro, S. (2010). Estimating QP and Motion Vectors in H.264/AVC Video from Decoded Pixels. *Proceedings of the 2nd ACM Workshop on Multimedia in Forensics, Security and Intelligence (MiFor '10)* (p. 89). New York: ACM Press.

Van Buskirk, E., & Liu, V. T. (2006). Digital Evidence: Challenging the Presumption of Reliability. *Journal of Digital Forensic Practice 1* (1), 19–26.

W3C JSON-LD Community Group. (n.d.). *JSON for Linking Data*. Retrieved July 26, 2019, from https://json-ld.org

Wang, S.-Y., Wang, O., Zhang, R., Owens, A., & Efros, A. A. (2020). CNN-Generated Images are Surprisingly Easy to Spot . . . for Now. *IEEE/CVF Conference on Computer Vision and Pattern Recognition (CVPR)* (pp. 8692–8701). IEEE.

Wang, W., & Farid, H. (2006). Exposing Digital Forgeries in Video by Detecting Double Quantization. *Proceedings of the 11th ACM Workshop on Multimedia and Security - MM&Sec '06* (p. 39). New York: ACM Press.

Wang, W., & Farid, H. (2007, September). Exposing Digital Forgeries in Interlaced and Deinterlaced Video. *IEEE Transactions on Information Forensics and Security 2* (3), 438–449.

Wang, Y., Uehara, T., & Sasaki, R. (2015). Fogcomputing: Issues and Challenges in Security and Forensics. *Proceedings - International Computer Software and Applications Conference 3*, 53–59.

Wheeler, D. A., & Larsen, G. N. (2003). *Techniques for Cyber Attack Attribution*. Institute for Defense Analyses (IDA). Retrieved from https://www.researchgate.net/publication/235170094_Techniques_for_Cyber_Attack_Attribution.

Williams, H. J., & Blum, I. (2018). *Defining Second Generation Open Source Intelligence (OSINT) for the Defense Enterprise*. RAND Corporation. Retrieved from https://www.rand.org/pubs/research_reports/RR1964.html

Winter, P., Ensafi, R., Loesing, K., & Feamster, N. (2016). Identifying and Characterizing Sybils in the Tor Network. *USENIX Security*.

Winter, T., Thubert, P., Brandt, A., Hui, J., Kelsey, R., Levis, P., Vasseur, JP., Pister, K., Struik, R., & Alexander, R. (2012). *RFC 6550: RPL: IPv6 Routing Protocol for Low-Power and Lossy Networks*. IETF.

World Wide Web Consortium. (n.d.). *Web of Things at W3C*. Retrieved June 30, 2019, from https://www.w3.org/WoT/

Yam, D., Wang, R., Zhou, J., Jin, C., & Wang, Z. (2018). Compression History Detection for MP3 Audio. *KSII Transactions on Internet and Information Systems 12* (2), 662–675.

Yang, R., Shi, Y.-Q., & Huang, J. (n.d.). Defeating Fake-Quality MP3. *Proceedings of the 11th ACM Workshop on Multimedia and Security* (pp. 117–124). ACM.

Z v Finland, Application no. 22009/93 (European Court of Human Rights (ECtHR) February 25, 1997).

Zakariah, M. A. (2018). Digital Multimedia Audio Forensics: Past, Present and Future. *Multimedia Tools and Applications 77*, 1009–1040.

Zapf, P. A., & Dror, I. E. (2017). Understanding and Mitigating Bias in Forensic Evaluation: Lessons from Forensic Science. *International Journal of Forensic Mental Health 16* (3), 227–238.

Zawoad, S., & Hasan, R. (2015). FAIoT: Towards Building a Forensics Aware Eco System for the Internet of Things. *Proceedings - 2015 IEEE International Conference on Services Computing, SCC 2015* (pp. 279–284).

Øverlier, L. (2007). Anonymity, Privacy and Hidden Services: Improving censorship-resistant publishing. Doctoral thesis, University Of Oslo.

Øverlier, L., & Syverson, P. (2006a, June). Valet Services: Improving Hidden Servers with a Personal Touch. In G. Danezis and P. Golle (Eds.), *Proceedings of the Sixth Workshop on Privacy Enhancing Technologies (PET 2006)*, LNCS 4258 (pp. 223–244). Cambridge: Springer-Verlag.

Øverlier, L., & Syverson, P. (2006b, May). Locating Hidden Servers. *Proceedings of the 2006 IEEE Symposium on Security and Privacy (S\&P'06)* (pp. 100–114). IEEE Computer Society. https://doi.org/10.1109/SP.2006.24

Årnes, A. (2018). *Digital Forensics*. Hoboken, NJ: John Wiley & Sons, Ltd.

Index